Nonlinear Estimation
Methods and Applications with Deterministic Sample Points

Nonlinear Estimation

Methods and Applications with Deterministic Sample Points

Shovan Bhaumik
Paresh Date

CRC Press
Taylor & Francis Group
Boca Raton London New York

CRC Press is an imprint of the
Taylor & Francis Group, an **Informa** business

A CHAPMAN & HALL BOOK

CRC Press
Taylor & Francis Group
6000 Broken Sound Parkway NW, Suite 300
Boca Raton, FL 33487-2742

International Standard Book Number-13: 978-0-8153-9432-7 (Hardback)

Library of Congress Control Number:2019946259

**Visit the Taylor & Francis Web site at
http://www.taylorandfrancis.com**

**and the CRC Press Web site at
http://www.crcpress.com**

To my parents

Shovan

To Bhagyashree

Paresh

Contents

Preface xiii

About the Authors xvii

Abbreviations xix

Symbol Description xxi

1 Introduction   1

  1.1 Nonlinear systems 2
    1.1.1 Continuous time state space model 2
    1.1.2 Discrete time state space model 3
  1.2 Discrete time systems with noises 5
    1.2.1 Solution of discrete time LTI system 6
    1.2.2 States as a Markov process 6
  1.3 Stochastic filtering problem 7
  1.4 Maximum likelihood and maximum a posterori estimate ... 8
    1.4.1 Maximum likelihood (ML) estimator 8
    1.4.2 Maximum a posteriori (MAP) estimate 9
  1.5 Bayesian framework of filtering 9
    1.5.1 Bayesian statistics 9
    1.5.2 Recursive Bayesian filtering: a conceptual solution ... 10
  1.6 Particle filter 12
    1.6.1 Importance sampling 12
    1.6.2 Resampling 15
  1.7 Gaussian filter 17
    1.7.1 Propagation of mean and covariance of a linear
        system 17
    1.7.2 Nonlinear filter with Gaussian approximations 19
  1.8 Performance measure 22
    1.8.1 When truth is known 22
    1.8.2 When truth is unknown 23
  1.9 A few applications 23
    1.9.1 Target tracking 23
    1.9.2 Navigation 24

1.9.3	Process control	24
1.9.4	Weather prediction	24
1.9.5	Estimating state-of-charge (SoC)	24
1.10	Prerequisites	25
1.11	Organization of chapters	25

2 The Kalman filter and the extended Kalman filter **27**

2.1	Linear Gaussian case (the Kalman filter)	27
2.1.1	Kalman filter: a brief history	27
2.1.2	Assumptions	28
2.1.3	Derivation	29
2.1.4	Properties: convergence and stability	31
2.1.5	Numerical issues	32
2.1.6	The information filter	33
2.1.7	Consistency of state estimators	34
2.1.8	Simulation example for the Kalman filter	35
2.1.9	MATLAB®-based filtering exercises	37
2.2	The extended Kalman filter (EKF)	38
2.2.1	Simulation example for the EKF	40
2.3	Important variants of the EKF	43
2.3.1	The iterated EKF (IEKF)	43
2.3.2	The second order EKF (SEKF)	45
2.3.3	Divided difference Kalman filter (DDKF)	45
2.3.4	MATLAB-based filtering exercises	49
2.4	Alternative approaches towards nonlinear filtering	49
2.5	Summary	50

3 Unscented Kalman filter **51**

3.1	Introduction	51
3.2	Sigma point generation	52
3.3	Basic UKF algorithm	54
3.3.1	Simulation example for the unscented Kalman filter	56
3.4	Important variants of the UKF	60
3.4.1	Spherical simplex unscented transformation	60
3.4.2	Sigma point filter with $4n+1$ points	61
3.4.3	MATLAB-based filtering exercises	64
3.5	Summary	64

4 Filters based on cubature and quadrature points **65**

4.1	Introduction	65
4.2	Spherical cubature rule of integration	66

4.3 Gauss-Laguerre rule of integration 67
4.4 Cubature Kalman filter 68
4.5 Cubature quadrature Kalman filter 70
    4.5.1 Calculation of cubature quadrature (CQ) points . . . 70
    4.5.2 CQKF algorithm 71
4.6 Square root cubature quadrature Kalman filter 75
4.7 High-degree (odd) cubature quadrature Kalman filter 77
    4.7.1 Approach 77
    4.7.2 High-degree cubature rule 77
    4.7.3 High-degree cubature quadrature rule 79
    4.7.4 Calculation of HDCQ points and weights 80
    4.7.5 Illustrations 80
    4.7.6 High-degree cubature quadrature Kalman filter 86
4.8 Simulation examples 87
    4.8.1 Problem 1 87
    4.8.2 Problem 2 91
4.9 Summary 92

5 Gauss-Hermite filter   **95**

5.1 Introduction 95
5.2 Gauss-Hermite rule of integration 96
    5.2.1 Single dimension 96
    5.2.2 Multidimensional integral 97
5.3 Sparse-grid Gauss-Hermite filter (SGHF) 99
    5.3.1 Smolyak's rule 100
5.4 Generation of points using moment matching method 104
5.5 Simulation examples 105
    5.5.1 Tracking an aircraft 105
5.6 Multiple sparse-grid Gauss-Hermite filter (MSGHF) 109
    5.6.1 State-space partitioning 109
    5.6.2 Bayesian filtering formulation for multiple
        approach 110
    5.6.3 Algorithm of MSGHF 111
    5.6.4 Simulation example 113
5.7 Summary 116

6 Gaussian sum filters   **117**

6.1 Introduction 117
6.2 Gaussian sum approximation 118
    6.2.1 Theoretical foundation 118
    6.2.2 Implementation 120
    6.2.3 Multidimensional systems 121
6.3 Gaussian sum filter 122

      6.3.1   Time update 122

      6.3.2   Measurement update 123

  6.4 Adaptive Gaussian sum filtering 124

  6.5 Simulation results 125

      6.5.1   Problem 1: Single dimensional nonlinear system 125

      6.5.2   RADAR target tracking problem 129

      6.5.3   Estimation of harmonics 133

  6.6 Summary 136

7 Quadrature filters with randomly delayed measurements   139

  7.1 Introduction 139

  7.2 Kalman filter for one step randomly delayed
measurements 140

  7.3 Nonlinear filters for one step randomly delayed
measurements 143

      7.3.1   Assumptions 144

      7.3.2   Measurement noise estimation 144

      7.3.3   State estimation 145

  7.4 Nonlinear filter for any arbitrary step randomly
delayed measurement 146

      7.4.1   Algorithm 153

  7.5 Simulation 154

  7.6 Summary 155

8 Continuous-discrete filtering   159

  8.1 Introduction 159

  8.2 Continuous time filtering 160

      8.2.1   Continuous filter for a linear Gaussian system 161

      8.2.2   Nonlinear continuous time system 167

           8.2.2.1   The extended Kalman-Bucy filter 167

  8.3 Continuous-discrete filtering 168

      8.3.1   Nonlinear continuous time process model 171

      8.3.2   Discretization of process model using
Runge-Kutta method 172

      8.3.3   Discretization using Ito-Taylor expansion of
order 1.5 172

      8.3.4   Continuous-discrete filter with deterministic
sample points 174

  8.4 Simulation examples 176

      8.4.1   Single dimensional filtering problem 176

      8.4.2   Estimation of harmonics 177

      8.4.3   RADAR target tracking problem 179

  8.5 Summary 186

9 Case studies 187

9.1 Introduction 187
9.2 Bearing only underwater target tracking problem 188
9.3 Problem formulation 189
   9.3.1 Tracking scenarios 190
9.4 Shifted Rayleigh filter (SRF) 191
9.5 Gaussian sum shifted Rayleigh filter (GS-SRF) 193
   9.5.1 Bearing density 194
9.6 Continuous-discrete shifted Rayleigh filter
   (CD-SRF) 194
   9.6.1 Time update of CD-SRF 196
9.7 Simulation results 196
   9.7.1 Filter initialization 199
   9.7.2 Performance criteria 201
   9.7.3 Performance analysis of Gaussian sum filters 201
   9.7.4 Performance analysis of continuous-discrete
       filters 211
9.8 Summary 215
9.9 Tracking of a ballistic target 216
9.10 Problem formulation 219
   9.10.1 Process model 219
       9.10.1.1 Process model in discrete domain 219
       9.10.1.2 Process model in continuous time
           domain 220
   9.10.2 Seeker measurement model 220
   9.10.3 Target acceleration model 223
9.11 Proportional navigation guidance (PNG) law 225
9.12 Simulation results 226
   9.12.1 Performance of adaptive Gaussian sum filters 228
   9.12.2 Performance of continuous-discrete filters 229
9.13 Conclusions 230

Bibliography 235

Index 251

Preface

This book deals with nonlinear state estimation. It is well known that, for a linear system and additive Gaussian noise, an optimal solution is available for the state estimation problem. This well known solution is known as the Kalman filter. However, if the systems are nonlinear, the posterior and the prior probability density functions (pdfs) are no longer Gaussian. For such systems, no optimal solution is available in general. The primitive approach is to linearize the system and apply the Kalman filter. The method is known as the extended Kalman filter (EKF). However, the estimate fails to converge in many cases if the system is highly nonlinear. To overcome the limitations associated with the extended Kalman filter, many techniques are proposed. All the post-EKF techniques could be divided into two categories, namely (i) the estimation with probabilistic sample points and (ii) the estimation with deterministic sample points. The probabilistic sample point methods approximately reconstruct the posterior and the prior pdfs with the help of many points in the state space (also known as *particles*) sampled from an appropriate probability distribution and their associated probability weights. On the other hand, deterministic sample point techniques approximate the posterior and the prior pdfs with a multidimensional Gaussian distribution and calculate the mean and covariance with a few wisely chosen points and weights. For this reason, they are also called Gaussian filters. They are popular in real time applications due to their ease of implementation and faster execution, when compared to the techniques based on probabilistic sample points.

There are good books on filtering with probabilistic sample points, i.e., particle filtering. However, the same is not true for approximate Gaussian filters. Moreover, over the last few years there is considerable development on the said topic. This motivates us to write a book which presents a complete coverage of the Bayesian estimation with deterministic sample points. The purpose of the book is to educate the readers about all the available Gaussian estimators. Learning of various available methods becomes essential for a designer as in filtering there is no 'holy grail' which will always provide the best result irrespective of the problems encountered. In other words, the best choice of estimator is highly problem specific.

There are prerequisites to understand the material presented in this book. These include (i) understanding of linear algebra and linear systems (ii) Bayesian probability theory, (iii) state space analysis. Assuming the readers are exposed to the above prerequisites, the book starts with the conceptual solution of the nonlinear estimation problems and describes all the Gaussian filters in depth with rigorous mathematical analysis.

The style of writing is suitable for engineers and scientists. The material of the book is presented with the emphasis on key ideas, underlying assumptions behind them, algorithms, and properties. In this book, readers will get a comprehensive idea and understanding about the approximate solutions of the nonlinear estimation problem. The designers, who want to implement the filters, will benefit from the algorithms, flow charts and MATLAB® code provided in the book. Rigorous, state of the art mathematical treatment will also be provided where relevant, for the analyst who wants to analyze the algorithm in depth for deeper understanding and further contribution. Further, beginners can verify their understanding with the help of numerical illustrations and MATLAB codes.

The book contains nine chapters. It starts with the formulation of the state estimation problem and the conceptual solution of it. Chapter 2 provides an optimal solution of the problem for a linear system and Gaussian noises. Further, it provides a detailed overview of several nonlinear estimators available in the literature. The next chapter deals with the unscented Kalman filter. Chapters 4 and 5 describe cubature and quadrature based Kalman filters, the Gauss-Hermite filter and their variants respectively. The next chapter presents the Gaussian sum filter, where the prior and the posterior pdfs are approximated with the weighted sum of several Gaussian pdfs. Chapter 7 considers the problem where measurements are randomly delayed. Such filters are finding more and more applications in networked control systems. Chapter 8 presents an estimation method for the continuous-discrete system. Such systems naturally arise because process equations are in continuous time domain as they are modeled from physical laws and the measurement equations are in discrete time domain as they arrive from the sampled sensor measurement. Finally, in the last chapter two case studies namely (i) bearing only underwater target tracking and (ii) tracking a ballistic target on reentry have been considered. All the Gaussian filters are applied to them and results are compared. Readers are suggested to start with the first two chapters because the rest of the book depends on them. Next, the reader can either read all the chapters from 3 to 6, or any of them (based on necessity). In other words, Chapters 3-6 are not dependent on one another. However, to read Chapters 7 to 9 understanding of the previous chapters is required.

This book is an outcome of many years of our research work, which was carried out with the active participation of our PhD students. We are thankful to them. Particularly, we would like to express our special appreciation and thanks to Dr Rahul Radhakrishnan and Dr Abhinoy Kumar Singh. Further, we thank anonymous reviewers, who reviewed our book proposal, for their constructive comments which help to uplift the quality of the book. We acknowledge the help of Mr Rajesh Kumar for drawing some of the figures included in the book. Finally, we would like to acknowledge with gratitude, the support and love of our families who all help us to move forward and this book would not have been possible without them.

We hope that the book will make significant contribution in the literature of Bayesian estimation and the readers will appreciate the effort. Further, it is anticipated that the book will open up many new avenues of both theoretical and applied research in various fields of science and technology.

Shovan Bhaumik
Paresh Date

MATLAB® and Simulink® are the registered trademark of The MathWorks, Inc. For product information, please contact:

The MathWorks, Inc.
3 Apple Hill Drive
Natick, MA, 01760-2098 USA
Tel: 508-647-7000
Fax: 508-647-7001
E-mail: info@mathworks.com
Web: https://www.mathworks.com

About the Authors

Dr. Shovan Bhaumik was born in Kolkata, India, in 1978. He received the B.Sc. degree in Physics in 1999 from Calcutta University, Kolkata, India, the B.Tech degree in Instrumentation and Electronics Engineering in 2002, the Master of Control System Engineering degree in 2004, and the PhD degree in Electrical Engineering in 2009, all from Jadavpur University, Kolkata, India.

He is currently Associate Professor of Electrical Engineering at Indian Institute of Technology Patna, India. From May 2007 to June 2009, he was a Research Engineer, at GE Global Research, John F Welch Technology Centre, Bangalore, India. From July 2009 to March 2017, he was an Assistant Professor of Electrical Engineering at Indian Institute of Technology Patna.

Shovan Bhaumik's research interests include nonlinear estimation, statistical signal processing, aerospace and underwater target tracking, and networked control systems. He has published more than 20 papers in refereed international journals. He is a holder of the Young Faculty Research Fellowship (YFRF) award from the Ministry of Electronics and Information Technology, MeitY, Government of India.

Dr. Paresh Date was born in 1971 in Mumbai, India. He completed his B.E. in Electronics and Telecommunication in 1993 from Pune University, India, his M.Tech. in Control and Instrumentation in 1995 from the Indian Institute of Technology Bombay (Mumbai), India and his doctoral studies in engineering at Cambridge University in 2001. His studies were funded by the Cambridge Commonwealth Trust (under the Cambridge Nehru Fellowship) and the CVCP, UK. He worked as a postdoctoral researcher at the University of Cambridge from 2000 to 2002. He joined Brunel University London in 2002, where he is currently a senior lecturer and Director of Research in the Department of Mathematics.

Dr. Date's principal research interests include filtering and its applications, especially in financial mathematics. He has published more than 50 refereed papers and supervised 10 PhD students to completion as their principal supervisor. His research has been funded by grants from the Engineering and

Physical Sciences Research Council, UK, from charitable bodies such as the London Mathematical Society, the Royal Society and from the industry. He has held visiting positions at universities in Australia, Canada and India. He is a Fellow of the Institute of Mathematics and its Applications and an Associate Editor for the *IMA Journal of Management Mathematics*.

Abbreviations

AGS	Adaptive Gaussian sum
AGSF	Adaptive Gaussian sum filter
ATC	Air traffic control
BOT	Bearings-only tracking
CKF	Cubature Kalman filter
CD	Continuous-discrete
CD-CKF	Continuous-discrete cubature Kalman filter
CD-CQKF	Continuous-discrete cubature quadrature Kalman filter
CD-EKF	Continuous-discrete extended Kalman filter
CDF	Central difference filter
CD-GHF	Continuous-discrete Gauss-Hermite filter
CD-NUKF	Continuous-discrete new unscented Kalman filter
CD-SGHF	Continuous-discrete sparse-grid Gauss-Hermite filter
CD-SRF	Continuous-discrete shifted Rayleigh filter
CD-UKF	Continuous-discrete unscented Kalman filter
CQKF	Cubature quadrature Kalman fiter
CQKF-RD	Cubature Kalman filter for random delay
CRLB	Cramer Rao lower bound
DDF	Divided difference filter
EKF	Extended Kalman filter
GHF	Gauss-Hermite filter
GPS	Global positioning system
GSF	Gaussian sum filter
GS-EKF	Gaussian sum extended Kalman filter
GS-CKF	Gaussian sum cubature Kalman filter
GS-CQKF	Gaussian sum cubature quadrature Kalman filter
GS-GHF	Gaussian sum Gauss-Hermite filter
GS-NUKF	Gaussian sum new unscented Kalman filter
GS-SGHF	Gaussian sum sparse-grid Gauss-Hermite filter
GS-SRF	Gaussian sum shifted Rayleigh filter
GS-UKF	Gaussian sum unscented Kalman filter
HDCKF	High-degree cubature Kalman filter
HDCQKF	High-degree cubature quadrature Kalman filter
IEKF	Iterated extended Kalman filter
MSGHF	Multiple sparse-grid Gauss-Hermite filter
NEES	Normalized estimation error squared

NIS	Normalized innovation squared
NUKF	New unscented Kalman filter
pdf	probability density function
PF	Particle filter
PNG	Proportional navigation guidance
RADAR	Radio detection and ranging
RMSE	Root mean square error
RSUKF	Risk sensitive unscented Kalman filter
SDC	State dependent coefficient
SEKF	Second order extended Kalman filter
SGHF	Sparse-grid Gauss-Hermite filter
SGQ	Sparse-grid quadrature
SIS	Sequential importance sampling
SIS-PF	Sequential importance sampling particle filter
SRCKF	Square-root cubature Kalman filter
SRCQKF	Square-root cubature quadrature Kalman filter
SRF	Shifted Rayleigh filter
SRNUKF	Square-root new unscented Kalman filter
SRUKF	Square-root unscented Kalman filter
SoC	State of charge
SONAR	Sound navigation and ranging
TMA	Target motion analysis
UKF	Unscented Kalman filter

Symbol Description

$\mathcal{X}$	State vector	k	Time step
$\mathcal{Y}$	Measurement	$\mathbb{R}^n$	Real space of dimension n
$\mathcal{U}$	Input	$\mathcal{N}$	Normal distribution
Q	Process noise covariance	$\phi(\cdot)$	Process nonlinear function
R	Measurement noise covariance	$\gamma(\cdot)$	Measurement nonlinear function
$\mathcal{Z}$	Delayed measurement	K	Kalman gain
p	Probability density function	A	System matrix
$\mathbb{E}$	Expectation	C	Measurement matrix
Σ	Error covariance matrix	$\bigcup$	Union of sample points
$\Sigma^{\mathcal{X}\mathcal{X}}$	Error covariance matrix of states	p	Latency parameter of delay
$\Sigma^{\mathcal{Y}\mathcal{Y}}$	Error covariance matrix of measurement	$\mathfrak{R}_i$	i^{th} order Runge-Kutta operator
$\Sigma^{\mathcal{X}\mathcal{Y}}$	Cross covariance of state and measurement	∇_{γ_k}	Jacobian of $\gamma_k(\cdot)$
η	Process noise	S	Cholesky factor
v	Measurement noise	$\mathcal{U}$	Input vector
$\hat{\mathcal{X}}$	Expectation of state vector	B	Input matrix
		$\mathbb{P}$	Probability

Chapter 1

Introduction

1.1	Nonlinear systems	2
	1.1.1 Continuous time state space model	2
	1.1.2 Discrete time state space model	3
1.2	Discrete time systems with noises	5
	1.2.1 Solution of discrete time LTI system	6
	1.2.2 States as a Markov process	6
1.3	Stochastic filtering problem	7
1.4	Maximum likelihood and maximum a posterori estimate	8
	1.4.1 Maximum likelihood (ML) estimator	8
	1.4.2 Maximum a posteriori (MAP) estimate	9
1.5	Bayesian framework of filtering	9
	1.5.1 Bayesian statistics	9
	1.5.2 Recursive Bayesian filtering: a conceptual solution	10
1.6	Particle filter	12
	1.6.1 Importance sampling	12
	1.6.2 Resampling	15
1.7	Gaussian filter	17
	1.7.1 Propagation of mean and covariance of a linear system	17
	1.7.2 Nonlinear filter with Gaussian approximations	19
1.8	Performance measure	22
	1.8.1 When truth is known	22
	1.8.2 When truth is unknown	23
1.9	A few applications	23
	1.9.1 Target tracking	23
	1.9.2 Navigation	24
	1.9.3 Process control	24
	1.9.4 Weather prediction	24
	1.9.5 Estimating state-of-charge (SoC)	24
1.10	Prerequisites	25
1.11	Organization of chapters	25

1.1 Nonlinear systems

All of us know that the dynamical behavior of a system can be described with mathematical models. For example, the flight path of an airplane subjected to engine thrust and under certain deflection of control surfaces, or the position and velocity of an underwater target can be described using mathematical models. The dynamics of a system is generally described with the help of a set of first order differential equations. Such a representation of a system is known as a state space model. The variables (which generally represent physical parameters) are known as states. Higher order differential equation models can often be re-formulated as augmented, first order differential equation models.

1.1.1 Continuous time state space model

From the above discussion, we see that a system can be represented with a set of variables (known as states) and the input output relationship can be represented with a set of first order differential equations. In mathematical language, $\mathcal{X} = [x_1 \ x_2 \ \cdots \ x_n]^T$ represents a state vector in n dimensional real space. In general, in continuous time a dynamic system can be represented by equations of the form [3]

$$\dot{x}_i = f_i(x_1, x_2, \cdots, x_n, u_1, u_2, \cdots, u_m, t), \qquad i = 1, \cdots, n, \qquad (1.1)$$

where $u_i, i = 1, \cdots, m$ denotes the external (or exogenous) inputs to the system. Output of the system is obtained from sensor measurements which can be represented as

$$y_i = g_i(x_1, x_2, \cdots, x_n, u_1, u_2, \cdots, u_m, t), \qquad i = 1, \cdots, p. \qquad (1.2)$$

f_i and g_i are nonlinear real valued functions of state, input and time. A complete description of such a system could be obtained if the Eqs. (1.1) and (1.2) are known along with a set of initial conditions of state vector. In a compact form and with matrix-vector notation, the above two equations can be represented as

$$\dot{\mathcal{X}} = f(\mathcal{X}, \mathcal{U}, t), \qquad (1.3)$$

and

$$\mathcal{Y} = g(\mathcal{X}, \mathcal{U}, t), \qquad (1.4)$$

where $f = [f_1 \ f_2 \ \cdots \ f_n]^T$, $g = [g_1 \ g_2 \ \cdots \ g_p]^T$, $\mathcal{U} = [u_1 \ u_2 \ \cdots \ u_m]^T$. Moreover, $\mathcal{X} \in \mathbb{R}^n$, $\mathcal{Y} \in \mathbb{R}^p$, and $\mathcal{U} \in \mathbb{R}^m$.

Important special cases of Eqs. (1.3), (1.4) are the linear time varying state and output equations given by

$$\dot{\mathcal{X}} = A(t)\mathcal{X} + B(t)\mathcal{U}, \qquad (1.5)$$

and

$$\mathcal{Y} = C(t)\mathcal{X} + D(t)\mathcal{U}, \tag{1.6}$$

where $A(t) \in \mathbb{R}^{n \times n}$, $B(t) \in \mathbb{R}^{n \times m}$, $C(t) \in \mathbb{R}^{p \times n}$, and $D(t) \in \mathbb{R}^{p \times m}$ respectively.

Now, if the system is time invariant, the state and output equations become

$$\dot{\mathcal{X}} = A\mathcal{X} + B\mathcal{U}, \tag{1.7}$$

and

$$\mathcal{Y} = C\mathcal{X} + D\mathcal{U}, \tag{1.8}$$

where A, B, C, D are real matrices with appropriate dimensions. The process is a first order differential equation and can be solved to obtain $\mathcal{X}(t)$ if initial condition is known [88].

1.1.2 Discrete time state space model

The state vector of a dynamic system in a discrete time domain can be represented as $\mathcal{X}_k = [x_{1,k}\ x_{2,k}\ \cdots\ x_{n,k}]^T$, where $\mathcal{X}_k \in \mathbb{R}^n$, and $k \in 0, 1, \cdots, N$, is the step count. When the step count is multiplied by sampling time, T, i.e., kT, we receive a time unit. Similar to Eqs. (1.1) and (1.2), in a discrete time nonlinear system the process and measurement equations can be written as

$$x_{i,k+1} = \phi_{i,k}(x_{1,k}, x_{2,k}, \cdots, x_{n,k}, u_{1,k}, u_{2,k}, \cdots, u_{m,k}, k), \qquad i = 1, \cdots, n, \tag{1.9}$$

where $u_{i,k}, i = 1, \cdots, m$ denotes the ith input of the system at any time step k. The output of the system which is generally sensor measurements can be represented as

$$y_{i,k} = \gamma_{i,k}(x_{1,k}, x_{2,k}, \cdots, x_{n,k}, u_{1,k}, u_{2,k}, \cdots, u_{m,k}, k), \qquad i = 1, \cdots, p. \tag{1.10}$$

In a compact form and with vector-matrix notation, the above two equations can be represented as

$$\mathcal{X}_{k+1} = \phi_k(\mathcal{X}_k, \mathcal{U}_k, k), \tag{1.11}$$

and

$$\mathcal{Y}_k = \gamma_k(\mathcal{X}_k, \mathcal{U}_k, k), \tag{1.12}$$

where $\phi_k(\cdot)$ and $\gamma_k(\cdot)$ are arbitrary nonlinear functions, and $\mathcal{X}_k \in \mathbb{R}^n$, $\mathcal{Y}_k \in \mathbb{R}^p$, and $\mathcal{U}_k \in \mathbb{R}^m$.

For a linear time varying discrete system the process and measurement equations become

$$\mathcal{X}_{k+1} = A_k \mathcal{X}_k + B_k \mathcal{U}_k, \tag{1.13}$$

and

$$\mathcal{Y}_k = C_k \mathcal{X}_k + D_k \mathcal{U}_k, \tag{1.14}$$

where A_k, B_k, C_k, D_k are real matrices with appropriate dimensions. For a time invariant linear system the above two equations can further be written as

$$\mathcal{X}_{k+1} = A\mathcal{X}_k + B\mathcal{U}_k, \tag{1.15}$$

and

$$\mathcal{Y}_k = C\mathcal{X}_k + D\mathcal{U}_k. \tag{1.16}$$

The Eq. (1.15) is a difference equation and can be solved recursively if the initial state vector is known.

Very often, in filtering literature, the term $\mathcal{U}_k$ is dropped keeping in mind that it can be incorporated with the final algorithm very easily even if we do not consider it during formulation. Without loss of generality, from now onwards we write general process and measurement equations as

$$\mathcal{X}_{k+1} = \phi_k(\mathcal{X}_k, k), \tag{1.17}$$

and

$$\mathcal{Y}_k = \gamma_k(\mathcal{X}_k, k). \tag{1.18}$$

The propagation of state and measurement expressed by the above two equations could be represented by Figure 1.1.

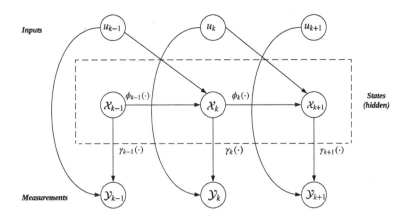

FIGURE 1.1: Graphical representation of evolution of state and measurement.

1.2 Discrete time systems with noises

The process equation models a system, generally using the law of physics. Most often, the model is not absolutely accurate. This inaccuracy is incurred in the system due to an error in modeling or the absence of full knowledge about the system and its input. Let us take an example of tracking a maneuvering enemy aircraft [141]. If we model the system with a constant turn rate, but actually it maneuvers with a variable turn rate, the developed model would be inaccurate. A popular method of handling this inaccuracy in modeling is to incorporate random process, which follows certain probability distribution, with the process model. The random number incorporated to compensate the modeling error is known as process noise. The process noise could be additive or multiplicative [1, 183]. Modeling the process uncertainty with additive noise is most popular and mathematically it could be written as

$$\mathcal{X}_{k+1} = \phi_k(\mathcal{X}_k, k) + \eta_k, \tag{1.19}$$

where η_k is the process noise. It should be noted that a caveat is necessary on the term process noise which may be misleading. The η_k is not actually noise; rather it is process excitation. However, the term process noise is very popular in the literature and throughout the book we shall call it by the same.

Some or all of the states of the system, or a function thereof, are measured by a set of sensors. We denote these measurements as outputs of the system, and the relationship between the state vector and the outputs is governed by the measurement equation of the system. The relation between the state and output is governed by the measurement equation of the system. For additive sensor noise, the measurement equation becomes

$$\mathcal{Y}_k = \gamma_k(\mathcal{X}_k, k) + v_k, \tag{1.20}$$

where v_k is sensor noise and is characterized with a probability density function.

If the process and measurement noise are not additive, the process and measurement equations in general could be written as,

$$\mathcal{X}_{k+1} = \phi_k(\mathcal{X}_k, k, \eta_k),$$

and

$$\mathcal{Y}_k = \gamma_k(\mathcal{X}_k, k, v_k).$$

In what follows, we will assume that the systems have additive process and measurement noise. In addition, we will also assume that the process and measurement noises are stationary white signals with known statistics [180]. Generally, they are described with a multidimensional Gaussian probability density function (pdf) with zero mean and appropriate covariance.

1.2.1   Solution of discrete time LTI system

No analytical solution is available to solve a nonlinear discrete time system with additive noise. However, for a linear system and with additive noise, the probability density function of the states can be evaluated at any particular instant of time if the initial states are given. A linear state space equation with additive noise can be written as

$$\mathcal{X}_{k+1} = A_k \mathcal{X}_k + \eta_k,$$

and

$$\mathcal{Y}_k = C_k \mathcal{X}_k + v_k.$$

From the above equation we can write,

$$\begin{aligned}
\mathcal{X}_{k+1} &= A_k \mathcal{X}_k + \eta_k \\
&= A_k A_{k-1} \mathcal{X}_{k-1} + A_k \eta_{k-1} + \eta_k
\end{aligned}$$

$$\vdots \tag{1.21}$$

$$= \Big[\prod_{j=0}^{k} A_{k-j} \Big] \mathcal{X}_0 + \sum_{i=0}^{k} \Big[\prod_{j=0}^{k-i-1} A_{k-j} \Big] \eta_i$$

If the upper index of the product term is lower than the lower index, the result is taken as an identity matrix. The matrix $\prod_{j=0}^{k} A_{k-j}$ is known as a state transition matrix. For a discrete time LTI system, the above equation becomes

$$\mathcal{X}_{k+1} = A^k \mathcal{X}_0 + \sum_{i=0}^{k} A^{k-i} \eta_i. \tag{1.22}$$

For a nonlinear discrete state space equation, the evolution of state could be obtained sequentially by passing the previous state, $\mathcal{X}_k$, through the nonlinear function $\phi(\mathcal{X}_k, k)$ and adding a noise sequence.

1.2.2   States as a Markov process

From Eq. (1.21), the expression of state at any instant k, evolved from any instant $l < k$, can be expressed as

$$\mathcal{X}_k = \Big[\prod_{j=0}^{k-l-1} A_{k-1-j} \Big] \mathcal{X}_l + \sum_{i=l}^{k-1} \Big[\prod_{j=0}^{k-i-2} A_{k-1-j} \Big] \eta_i. \tag{1.23}$$

The state sequence $\mathcal{X}_{0:l}$ (where $\mathcal{X}_{0:l}$ means $\{\mathcal{X}_0, \mathcal{X}_1, \cdots, \mathcal{X}_l\}$) only depends on $\eta_{0:l-1}$. The noise sequence, $\eta_{1:l-1}$ is independent of $\mathcal{X}_{0:l}$. So, we could write $p(\mathcal{X}_k|\mathcal{X}_{0:l}) = p(\mathcal{X}_k|\mathcal{X}_l)$, which, in turn, means that the state vector is a Markov sequence. It should be noted that the above discussion holds true when the process noise is white, i.e., completely unpredictable [13]. If the noise is colored, the statement made above would not hold true as the state prior up to time l could be used to predict the process noise sequence incorporated during $k = l, \cdots, k - 1$.

1.3 Stochastic filtering problem

We see that Eq. (1.19), mentioned earlier is a stochastic difference equation. At each point of time, the state vector is characterized by a multidimensional probability density function. The objective of state estimation is to compute the full joint probability distribution of states at each time steps. Computing the full joint distribution of the states at all time steps is computationally inefficient in real life implementation. So, most of the time marginal distributions of states are considered. Three different types of state estimations may be done as described below [137]:

- *Filtering:* Filtering is an operation that involves the determination of the pdf of states at any time step k, given that current and previous measurements are received. Mathematically, filtering is the determination of $p(\mathcal{X}_k|\mathcal{Y}_{1:k})$, where $\mathcal{Y}_{1:k} = \{\mathcal{Y}_1, \mathcal{Y}_2, \cdots, \mathcal{Y}_k\}$.

- *Prediction:* Prediction is an priori form of estimation. Its aim is to compute marginal distributions of $\mathcal{X}_{k+n}$, (where $n = 1, 2, \cdots$), n step in advance of current measurement. Mathematically, prediction is the determination of $p(\mathcal{X}_{k+n}|\mathcal{Y}_{1:k})$. Unless specified otherwise, prediction is generally referred to as one step ahead estimation i.e., $n = 1$.

- *Smoothing:* Smoothing is an posteriori estimation of marginal distribution of $\mathcal{X}_k$, when measurements are received up to N step, where $N > k$. In other words, smoothing is the computation of $p(\mathcal{X}_k|\mathcal{Y}_{1:N})$.

The concept of filtering, prediction and smoothing is illustrated in Figure 1.2. Throughout the book, we shall consider only the filtering problem for a nonlinear system with additive white Gaussian noise as described in Eqs. (1.19) and (1.20). Further, a marginalized distribution of state will be considered. It must be noted that Eq. (1.19) characterizes the state transition density $p(\mathcal{X}_k|\mathcal{X}_{k-1})$ and Eq. (1.20) expresses the measurement density $p(\mathcal{Y}_k|\mathcal{X}_k)$.

As discussed above, the objective of the filtering is to estimate the pdf of the present state at time step k, given that the observation up to the kth step has been received i.e., $p(\mathcal{X}_k|\mathcal{Y}_{1:k})$, which is essentially the posterior density function. However, in many practical problems, the user wants a single vector as the estimated value of the state vector rather than a description of pdf. In such cases, the mean of the pdf is declared as the point estimate of the states.

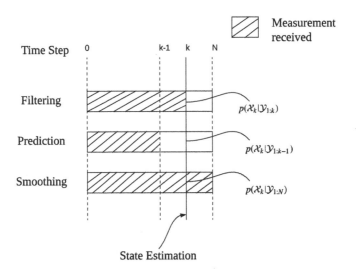

FIGURE 1.2: Illustration of filtering, prediction and smoothing.

1.4 Maximum likelihood and maximum a posterori estimate

1.4.1 Maximum likelihood (ML) estimator

In the non-Bayesian approach (we shall discuss what is the Bayesian approach very soon) there is no a priori pdf. So Bayes' formula could not be used. In this case, the pdf of the measurement conditioned on state or parameter, known as likelihood, can be obtained. So the likelihood function is

$$\Lambda_k(\mathcal{X}_k) \triangleq p(\mathcal{Y}_k|\mathcal{X}_k), \tag{1.24}$$

which measures how likely the state value is given the observations. The states or parameters can be estimated by maximizing the likelihood function. Thus the maximum likelihood estimator is

$$\hat{\mathcal{X}}_k^{ML} = \underset{\mathcal{X}_k}{arg\ max}\ \Lambda_k(\mathcal{X}_k) = \underset{\mathcal{X}_k}{arg\ max}\ p(\mathcal{Y}_k|\mathcal{X}_k). \tag{1.25}$$

The above optimization problem needs to be solved to obtain the maximum likelihood estimate.

1.4.2 Maximum a posteriori (MAP) estimate

The maximum a posteriori (MAP) estimator maximizes the posterior pdf. So the MAP estimate is

$$\hat{\mathcal{X}}_k^{MAP} = arg\ \underset{\mathcal{X}_k}{max}\ p(\mathcal{X}_k|\mathcal{Y}_k). \tag{1.26}$$

With the help of Bayes' theorem which says

$$p(\mathcal{X}_k|\mathcal{Y}_k) = \frac{p(\mathcal{Y}_k|\mathcal{X}_k)p(\mathcal{X}_k)}{p(\mathcal{Y}_k)}, \tag{1.27}$$

the MAP estimate becomes

$$\hat{\mathcal{X}}_k^{MAP} = arg\ \underset{\mathcal{X}_k}{max}\ [p(\mathcal{Y}_k|\mathcal{X}_k)p(\mathcal{X}_k)]. \tag{1.28}$$

Here, we drop the term $p(\mathcal{Y}_k)$ because it is a normalization constant and is irrelevant for a maximization problem.

1.5 Bayesian framework of filtering

1.5.1 Bayesian statistics

Bayesian theory is a branch of probability theory that allows us to model an uncertainty about the outcome of interest by incorporating prior knowledge about the system and observational evidence with the help of Bayes' theorem. The Bayesian method which interprets the probability as a conditional measure of uncertainty, is a very popular and useful tool as far as practical applications are concerned. In the filtering context, the Bayesian approach uses the prior pdf and the measurement knowledge to infer the conditional probability of states.

At the onset, we assume that (i) the states of a system follow a first order Markov process, so $p(\mathcal{X}_k|\mathcal{X}_{0:k-1}) = p(\mathcal{X}_k|\mathcal{X}_{k-1})$; (ii) the states are independent of the given measurements. From the Bayes' rule we have [76]

$$
\begin{aligned}
p(\mathcal{X}_k|\mathcal{Y}_{1:k}) &= \frac{p(\mathcal{Y}_{1:k}|\mathcal{X}_k)p(\mathcal{X}_k)}{p(\mathcal{Y}_{1:k})} \\
&= \frac{p(\mathcal{Y}_k,\mathcal{Y}_{1:k-1}|\mathcal{X}_k)p(\mathcal{X}_k)}{p(\mathcal{Y}_k,\mathcal{Y}_{1:k-1})} \\
&= \frac{p(\mathcal{Y}_k|\mathcal{Y}_{1:k-1},\mathcal{X}_k)p(\mathcal{Y}_{1:k-1}|\mathcal{X}_k)p(\mathcal{X}_k)}{p(\mathcal{Y}_k|\mathcal{Y}_{1:k-1})p(\mathcal{Y}_{1:k-1})} \\
&= \frac{p(\mathcal{Y}_k|\mathcal{Y}_{1:k-1},\mathcal{X}_k)p(\mathcal{X}_k|\mathcal{Y}_{1:k-1})p(\mathcal{Y}_{1:k-1})p(\mathcal{X}_k)}{p(\mathcal{Y}_k|\mathcal{Y}_{1:k-1})p(\mathcal{Y}_{1:k-1})p(\mathcal{X}_k)}.
\end{aligned}
$$

As we can write $p(\mathcal{Y}_k|\mathcal{Y}_{1:k-1}, \mathcal{X}_k) = p(\mathcal{Y}_k|\mathcal{X}_k)$, the above equation becomes

$$p(\mathcal{X}_k|\mathcal{Y}_{1:k}) = \frac{p(\mathcal{Y}_k|\mathcal{X}_k)p(\mathcal{X}_k|\mathcal{Y}_{1:k-1})}{p(\mathcal{Y}_k|\mathcal{Y}_{1:k-1})}. \tag{1.29}$$

Eq. (1.29) expresses the posterior pdf which consists of three terms, explained below.

- *Likelihood:* $p(\mathcal{Y}_k|\mathcal{X}_k)$ is the likelihood which essentially is determined from the measurement noise model of Eq. (1.20).

- *Prior:* $p(\mathcal{X}_k|\mathcal{Y}_{1:k-1})$ is defined as prior which can be obtained through the Chapman-Kolmogorov equation [8],

$$p(\mathcal{X}_k|\mathcal{Y}_{1:k-1}) = \int p(\mathcal{X}_k|\mathcal{X}_{k-1}, \mathcal{Y}_{1:k-1})p(\mathcal{X}_{k-1}|\mathcal{Y}_{1:k-1})d\mathcal{X}_{k-1}. \tag{1.30}$$

  As we assume the system to follow a first order Markov process, $p(\mathcal{X}_k|\mathcal{X}_{k-1}, \mathcal{Y}_{1:k-1}) = p(\mathcal{X}_k|\mathcal{X}_{k-1})$. Under such condition,

$$p(\mathcal{X}_k|\mathcal{Y}_{1:k-1}) = \int p(\mathcal{X}_k|\mathcal{X}_{k-1})p(\mathcal{X}_{k-1}|\mathcal{Y}_{1:k-1})d\mathcal{X}_{k-1}. \tag{1.31}$$

  The above equation is used to construct the prior pdf. $p(\mathcal{X}_k|\mathcal{X}_{k-1})$ could be determined from the process model of the system described in Eq. (1.19). If the likelihood and the prior could be calculated, the posterior pdf of states is estimated using Bayes' rule described in Eq. (1.29).

- *Normalization constant:* The denominator of the Eq. (1.29) is known as the normalization constant, or evidence and is expressed as

$$p(\mathcal{Y}_k|\mathcal{Y}_{1:k-1}) = \int p(\mathcal{Y}_k|\mathcal{X}_k)p(\mathcal{X}_k|\mathcal{Y}_{1:k-1})d\mathcal{X}_k. \tag{1.32}$$

1.5.2   Recursive Bayesian filtering: a conceptual solution

The key equations for state estimation are Eq. (1.31) and (1.29). It should also be noted that they are recursive in nature. This means that the state estimators designed with the help of Bayesian statistics, described above, have a potential to be implemented on a digital computer. The estimation may start with an initial pdf of the states, $p(\mathcal{X}_0|\mathcal{Y}_0)$ and may continue recursively. To summarize, the estimation of the posterior pdf of states can be performed with two steps:

- *Time update:* Time update, alternatively known as the prediction step, for a nonlinear filter is performed with the help of the process model described in Eq. (1.19) and the Chapman-Kolgomorov equation described in Eq. (1.31). The process model provides the state transition density, $p(\mathcal{X}_k|\mathcal{X}_{k-1})$ and the further Chapman-Kolgomorov equation determines the prior pdf $p(\mathcal{X}_k|\mathcal{Y}_{1:k-1})$.

- *Measurement update:* In this step (alternatively known as the correction step), the posterior pdf is calculated with the help of the prior pdf and likelihood. Likelihood is obtained from the measurement Eq. (1.20) and noise statistics. From Eq. (1.29), we know that

$$p(\mathcal{X}_k|\mathcal{Y}_{1:k}) \propto p(\mathcal{Y}_k|\mathcal{X}_k)p(\mathcal{X}_k|\mathcal{Y}_{1:k-1}). \tag{1.33}$$

The above equation is utilized to determine the posterior pdf of state.

Figure 1.3 shows the iterative time and measurement update process. Further, it should be kept in mind that the filtering strategy described above is only conceptual in nature. For a linear Gaussian system, $p(\mathcal{X}_k|\mathcal{X}_{k-1})$ and $p(\mathcal{X}_k|\mathcal{Y}_{1:k})$ will be Gaussian and a closed form solution is available. For any arbitrary nonlinear system, in general, no closed form solution is achievable. To estimate the states in such cases, the equations described above must be solved numerically with acceptable accuracy.

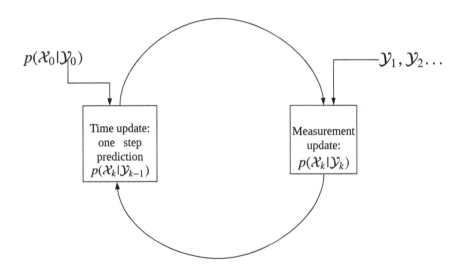

FIGURE 1.3: Recursive filtering in two steps.

1.6   Particle filter

We have seen earlier that our objective is to determine the prior and the posterior pdfs of the states. These pdfs can be represented with many points in space and their corresponding weights. As you may guess, this is a discretized representation of a continuous pdf. The representation can be made in two ways:

(i) The point mass description where the continuous pdfs are approximated with a set of support points whose weights depend on the location of the points. Definitely the weights are not equal.

(ii) Independent and identically distributed (iid) sample points where all the weights are the same but the points are denser where the probability density is more and vice versa. Both the point mass description and iid samples representation are illustrated in Figure 1.4.

In filtering literature, the points in state space are popularly known as particles from which the name particle filter (PF) [24] is derived. The key idea is to represent the posterior and prior pdf by a set of randomly sampled points and their weights. It is expected that as the number of particles becomes very large, the Monte Carlo characterization becomes an equivalent representation to the pdf under consideration (prior or posterior). As the particles are randomly sampled, the estimation method is also called filtering with random support points. This class of methods is also known variously as sequential importance sampling, sequential Monte Carlo method [41, 92], bootstrap filtering [57], condensation algorithm [19] etc.

As we mentioned earlier, the posterior pdf of state $p(\mathcal{X}_k|\mathcal{Y}_{1:k})$ is presented with N_s number of points in real space or particles denoted with $\mathcal{X}_k^i$, where $i = 1, \cdots, N_s$, and their corresponding weights w_k^i. So the posterior pdf could be represented as

$$p(\mathcal{X}_k|\mathcal{Y}_{1:k}) = \sum_{i=1}^{N_s} w_k^i \delta(\mathcal{X}_k - \mathcal{X}_k^i), \qquad (1.34)$$

where δ denotes the Dirac Delta function which are only defined at the location of particle. Similarly, the prior probability density function $p(\mathcal{X}_k|\mathcal{Y}_{1:k-1})$ can be represented. It should also be noted that the weights must be normalized, i.e., $\sum_{i=1}^{N_s} w_k^i = 1$. From the above equation we see that we need to determine the weight and obviously, we do not know the posterior pdf.

1.6.1   Importance sampling

A popular way is to determine the weights with the help of a powerful technique, known as importance sampling. Although we cannot draw samples from the posterior pdf, at each particle the $p(.)$ can be evaluated up to proportionality. Further, the posterior pdf is assigned with some guessed distribution, known as proposal density, from which the samples are generated easily. If we draw N_s samples from the proposal density, i.e., $\mathcal{X}_k^i \sim q(\mathcal{X}_k|\mathcal{Y}_{1:k})$, for $i = 1, \cdots, N_s$ the expression of weights in Eq. (1.34) becomes

$$w_k^i \propto \frac{p(\mathcal{X}_k^i|\mathcal{Y}_{1:k})}{q(\mathcal{X}_k^i|\mathcal{Y}_{1:k})}. \tag{1.35}$$

At each iteration, say k, we have particles from an earlier step posterior pdf, $p(\mathcal{X}_{k-1}|\mathcal{Y}_{1:k-1})$, and we want to draw samples from the present proposal density, $q(\mathcal{X}_k|\mathcal{Y}_{1:k})$. One way of doing this is by choosing an importance density

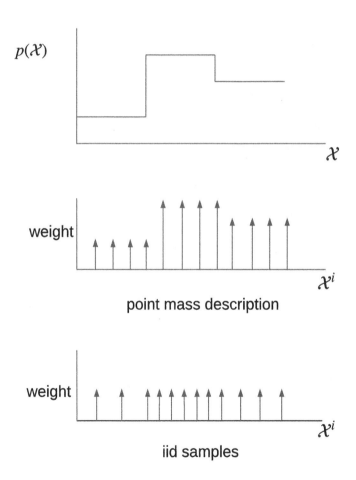

FIGURE 1.4: Point mass vs. iid sample approximation of a pdf.

which could be factorized as

$$q(\mathcal{X}_k|\mathcal{Y}_{1:k}) = q(\mathcal{X}_k|\mathcal{X}_{k-1}, \mathcal{Y}_{1:k})q(\mathcal{X}_{k-1}|\mathcal{Y}_{1:k-1}). \qquad (1.36)$$

Then the new particles $\mathcal{X}_k^i$ could be generated from the proposal, $q(\mathcal{X}_k|\mathcal{X}_{k-1})$.

Now we shall proceed to derive the weight update equation. Recall the Eq. (1.29) which can further be written as

$$
\begin{aligned}
p(\mathcal{X}_k|\mathcal{Y}_{1:k}) &= \frac{p(\mathcal{Y}_k|\mathcal{X}_k)p(\mathcal{X}_k|\mathcal{Y}_{1:k-1})}{p(\mathcal{Y}_k|\mathcal{Y}_{1:k-1})} \\
&= \frac{p(\mathcal{Y}_k|\mathcal{X}_k)p(\mathcal{X}_{k-1}|\mathcal{Y}_{1:k-1})p(\mathcal{X}_k|\mathcal{X}_{k-1})}{p(\mathcal{Y}_k|\mathcal{Y}_{1:k-1})} \qquad (1.37) \\
&\propto p(\mathcal{Y}_k|\mathcal{X}_k)p(\mathcal{X}_k|\mathcal{X}_{k-1})p(\mathcal{X}_{k-1}|\mathcal{Y}_{1:k-1}).
\end{aligned}
$$

With the help of Eq. (1.36) and (1.37) the Eq. (1.35) could be expressed as

$$
\begin{aligned}
w_k^i &\propto \frac{p(\mathcal{Y}_k|\mathcal{X}_k^i)p(\mathcal{X}_k^i|\mathcal{X}_{k-1}^i)p(\mathcal{X}_{k-1}^i|\mathcal{Y}_{1:k-1})}{q(\mathcal{X}_k^i|\mathcal{X}_{k-1}^i, \mathcal{Y}_{1:k})q(\mathcal{X}_{k-1}^i|\mathcal{Y}_{1:k-1})} \\
&= w_{k-1}^i \frac{p(\mathcal{Y}_k|\mathcal{X}_k^i)p(\mathcal{X}_k^i|\mathcal{X}_{k-1}^i)}{q(\mathcal{X}_k^i|\mathcal{X}_{k-1}^i, \mathcal{Y}_{1:k})}.
\end{aligned}
\qquad (1.38)
$$

The readers should note the following points:

(i) $p(\mathcal{Y}_k|\mathcal{X}_k^i)$ is known as likelihood which is the probability of obtaining a measurement corresponding to a particular particle. $p(\mathcal{X}_k^i|\mathcal{X}_{k-1}^i)$ is the transitional density.

(ii) $q(\mathcal{X}_k^i|\mathcal{X}_{k-1}^i, \mathcal{Y}_{1:k})$ is called the proposal density which is the practitioner's choice. The easiest way to choose proposal density is prior, i.e., $q(\mathcal{X}_k^i|\mathcal{X}_{k-1}^i, \mathcal{Y}_{1:k}) = p(\mathcal{X}_k^i|\mathcal{X}_{k-1}^i)$. Under such a choice, the weight update equation becomes $w_k^i \propto w_{k-1}^i p(\mathcal{Y}_k|\mathcal{X}_k^i)$.

(iii) A posterior probability density function obtained from any nonlinear filter such as the extended Kalman filter, the unscented Kalman filter [169], the cubature Kalman filter [159, 173], the Gaussian sum filter [95], or even the particle filter could be used as the proposal. It is reported that the accuracy of PF may be enhanced with the more accurate proposal.

(iv) At each step w_k^i should be normalized, i.e., $w_k^i = w_k^i / \sum_{i=1}^{N_s} w_k^i$, so that the particles and weights represent a probability mass function.

The methodology discussed above is commonly known as sequential importance sampling particle filter (SIS-PF) whose implementation at time step k is given in the Algorithm 1.

Algorithm 1 Sequential importance sampling particle filter

$$[\{\mathcal{X}_k^i, w_k^i\}_{i=1}^{N_s}] = \text{SIS}[\{\mathcal{X}_{k-1}^i, w_{k-1}^i\}_{i=1}^{N_s}, \mathcal{Y}_k]$$

- *for* $i = 1 : N_s$

    - Draw $\mathcal{X}_k^i \sim q(\mathcal{X}_k | \mathcal{X}_{k-1}^i, \mathcal{Y}_k)$
    - Compute weight $w_k^i = w_{k-1}^i \frac{p(\mathcal{Y}_k | \mathcal{X}_k^i) p(\mathcal{X}_k^i | \mathcal{X}_{k-1}^i)}{q(\mathcal{X}_k^i | \mathcal{X}_{k-1}^i, \mathcal{Y}_{1:k})}$

- *end for*

- Normalize the weights $w_k^i = w_k^i / \sum_{i=1}^{N_s} w_k^i$

1.6.2 Resampling

After a few iterations with the SIS algorithm, most of the particles will have a very small weight. This problem is known as weight degeneracy of the samples [106]. After a certain number of steps in the recursive algorithm, a large number of generated particles does not contribute to the posterior pdf and the computational effort is wasted. Further, it can be shown that the variance of the importance weight will only increase with time. Thus sample degeneracy in SIS algorithm is inevitable.

To get rid of such a problem, one may take a brute force approach by incorporating an increasingly large (potentially infinite) number of particles. However, that is not practically feasible to implement. Instead, it is beneficial to insert a resampling stage between two consecutive SIS recursions. Such a method is known as sequential importance sampling resampling (SISR). During the resampling stage, based on the existing particles and their weights, a new set of particles is generated and then equal weights are assigned to all of them.

It is argued that the resampling stage may not be necessary at each step. When the effective number of particles falls below a certain threshold limit, the resampling algorithm can be executed. An estimate of the effective sample size is given by [8]

$$N_{\text{eff}} = \frac{1}{\sum_{i=1}^{N_s} (w_k^i)^2}, \tag{1.39}$$

where w_k^i is normalized weights. Large N_{eff} represents high degeneracy. When $N_{eff} < N_T$ which is user defined, then we run the resampling algorithm. The SISR steps are mentioned in Algorithm 2. SISR filtering algorithm is depicted with the help of Figure 1.5.

In an earlier subsection, we have mentioned that the resampling is required to avoid the degeneracy problem of the particles. The basic idea behind resampling is to eliminate the particles that have small weights and repeat the particles with large weights. In other words, during the resampling process

---

Algorithm 2 Sequential importance sampling resampling filter

$$[\{\mathcal{X}_k^i, w_k^i\}_{i=1}^{N_s}] = \text{SISR}[\{\mathcal{X}_{k-1}^i, w_{k-1}^i\}_{i=1}^{N_s}, \mathcal{Y}_k]$$

- *for* $i = 1 : N_s$

    - Draw $\mathcal{X}_k^i \sim q(\mathcal{X}_k | \mathcal{X}_{k-1}^i, \mathcal{Y}_k)$
    - Compute weight $w_k^i = w_{k-1}^i \frac{p(\mathcal{Y}_k | \mathcal{X}_k^i) p(\mathcal{X}_k^i | \mathcal{X}_{k-1}^i)}{q(\mathcal{X}_k^i | \mathcal{X}_{k-1}^i, \mathcal{Y}_{1:k})}$

- *end for*

- Normalize the weights $w_k^i = w_k^i / \sum_{i=1}^{N_s} w_k^i$

- Compute the effective sample size as $N_{eff} = 1 / \sum_{i=1}^{N_s} (w_k^i)^2$

- Set the threshold (N_T) for effective sample size

- *if* $N_{eff} < N_T$

    - Resample using the importance weights
      $[\mathcal{X}_k^i, w_k^i\}_{i=1}^{N_s}] = \text{RESAMPLE}[\mathcal{X}_k^i, w_k^i\}_{i=1}^{N_s}]$

- *end if*

---

the particles with small weights are ignored whereas the particles with large weights are considered repeatedly. Further, all the new particles are assigned with equal weights. There are few practical disadvantages with the resampling process. Firstly, it limits the parallel implementation of the particle filtering algorithm; secondly as the particles with higher weights are taken repeatedly the particles after resampling become less diversified. This problem is known as particle impoverishment which is very severe for small process noise. In such a case, all the particles succumb to a single point leading to an erroneous estimate of the posterior pdf.

The most popular resampling strategy is systematic resampling [94, 8]. The pseudocode of systematic resampling [8] is described in Algorithm 3. Apart from systematic resampling various resampling schemes, such as multinomial resampling, stratified resampling [39], residual resampling [106] etc., are available in the literature. Interested readers are referred to [39] for various resampling strategies and their comparison.

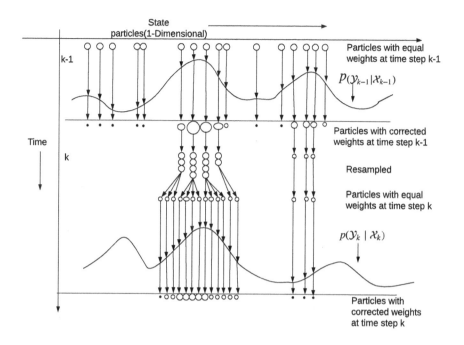

FIGURE 1.5: Sequential importance sampling resampling filter.

1.7 Gaussian filter

We have seen that the prior and the posterior pdf of a nonlinear system are in general non-Gaussian. However, many filtering methods assume the pdfs are Gaussian and characterize them with mean and covariance. Such filters are known as Gaussian filters. Sometimes they are also called deterministic sample point filters as they use a few deterministic points and their weights to determine mean and covariance of the pdfs. In this section we shall discuss the basic concept and formulation of such filters.

1.7.1 Propagation of mean and covariance of a linear system

If the system is linear, the process noise is Gaussian and the pdf of the current state is Gaussian, the updated pdf of the next time step is also Gaussian, but with a different mean and a different covariance. But the same is not true for a nonlinear system, the reason why no closed form estimator is available for such cases. The statement is illustrated in Figure 1.6.

Algorithm 3 Systematic resampling

$$[\{\mathcal{X}_k^j, w_k^j\}_{j=1}^{N_s}] = \text{RESAMPLE}[\{\mathcal{X}_k^i, w_k^i\}_{i=1}^{N_s}]$$

- Initialize the CDF: $c_1 = 0$

- Draw starting points of reference: $u_1 = U(0, 1/N_s)$

- For $i = 2 : N_s$
  - $c_i = c_{i-1} + w_k^i$

- End For

- Reset index:$i = 1$

- For $j = 1 : N_s$
  - $u_j = u_1 + (j-1)/N_s$
  - If $u_j > c_i$
  - $i = i + 1$
  - End If
  - Assign sample: $\mathcal{X}_k^j = \mathcal{X}_k^i$
  - Assign weight: $w_k^j = 1/N_s$

- End For

A linear discrete Gaussian system can be written as

$$\mathcal{X}_{k+1} = A_k \mathcal{X}_k + \eta_k, \tag{1.40}$$

where η_k is white Gaussian noise with mean $\bar{\eta}_k$ and covariance Q_k, i.e., $\eta_k \sim \mathcal{N}(\bar{\eta}_k, Q_k)$. Let us assume, at any step, k, the state variable follows $\mathcal{X}_k \sim \mathcal{N}(\hat{\mathcal{X}}_k, \Sigma^{\mathcal{X}\mathcal{X}})$, where $\Sigma^{\mathcal{X}\mathcal{X}}$ is the error covariance of states at time step k. Now the question is if $\mathcal{X}_k$ is propagated through the Eq. (1.40) what will be the mean and covariance of the pdf of states for the next time step $k+1$. The expressions for the mean and the error covariance of $\mathcal{X}_{k+1}$ at the next time step $k+1$ are

$$\hat{\mathcal{X}}_{k+1|k} = \mathbb{E}[A_k \mathcal{X}_{k|k} + \eta_k] = A_k \hat{\mathcal{X}}_{k|k} + \bar{\eta}_k, \tag{1.41}$$

and

$$
\begin{aligned}
\Sigma_{k+1|k}^{\mathcal{X}\mathcal{X}} &= \mathbb{E}[\mathcal{X}_{k+1} - \hat{\mathcal{X}}_{k+1|k}][\mathcal{X}_{k+1} - \hat{\mathcal{X}}_{k+1|k}]^T \\
&= \mathbb{E}[[A_k(\mathcal{X}_k - \hat{\mathcal{X}}_{k|k}) + (\eta_k - \bar{\eta}_k)][A_k(\mathcal{X}_k - \hat{\mathcal{X}}_{k|k}) + (\eta_k - \bar{\eta}_k)]^T] \\
&= \mathbb{E}[A_k(\mathcal{X}_k - \hat{\mathcal{X}}_{k|k})(\mathcal{X}_k - \hat{\mathcal{X}}_{k|k})^T A_k^T] + \mathbb{E}[A_k(\mathcal{X}_k - \hat{\mathcal{X}}_{k|k})(\eta_k - \bar{\eta}_k)^T] \\
&\quad + \mathbb{E}[(\eta_k - \bar{\eta}_k)(\mathcal{X}_k - \hat{\mathcal{X}}_{k|k})^T A_k^T] + \mathbb{E}[(\eta_k - \bar{\eta}_k)(\eta_k - \bar{\eta}_k)^T] \\
&= A_k \Sigma_{k|k}^{\mathcal{X}\mathcal{X}} A_k^T + Q_k.
\end{aligned}
\tag{1.42}
$$

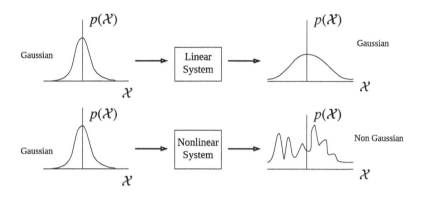

FIGURE 1.6: Gaussian pdf updated with linear and nonlinear systems.

The cross terms in the above equation vanish due to the whiteness of the process noise, which causes $\mathcal{X}_k$, being a linear combination of noise prior to k, to be independent of η_k [13].

1.7.2 Nonlinear filter with Gaussian approximations

Let us consider a nonlinear discrete system, expressed with

$$\mathcal{X}_{k+1} = \phi(\mathcal{X}_k, k) + \eta_k, \tag{1.43}$$

and

$$\mathcal{Y}_k = \gamma(\mathcal{X}_k, k) + v_k. \tag{1.44}$$

State vector $\mathcal{X}_k$, measurement $\mathcal{Y}_k$, process and measurement noise, η_k and v_k, belong to a real space of dimensions as mentioned in 1.1.2. Further assumptions on noise statistics are mentioned in the box below.

Assumption for Gaussian filters

Further, assume that the process and the measurement noises are white, zero mean Gaussian with the covariance Q_k and R_k respectively. The process and measurement noise sequences are also assumed to be uncorrelated to each other. Mathematically, $\eta_k \sim \mathcal{N}(0, Q_k)$, and $v_k \sim \mathcal{N}(0, R_k)$, which means

$$\mathbb{E}[\eta_k] = 0, \qquad \mathbb{E}[\eta_k \eta_j^T] = Q_k \delta_{kj} \qquad (1.45)$$

$$\mathbb{E}[v_k] = 0, \qquad \mathbb{E}[v_k v_j^T] = R_k \delta_{kj} \qquad (1.46)$$

Further, as the noises are uncorrelated,

$$\mathbb{E}[\eta_k v_j^T] = 0 \qquad \forall k, j. \qquad (1.47)$$

We know that if the system is nonlinear, even if the Gaussian pdf is updated with it, the resultant pdf becomes non-Gaussian. As discussed earlier, in many algorithms of nonlinear filtering, it is assumed that the prior and the posterior densities of states and the measurement density are all Gaussian and are characterized with the first two moments. The filters which rely on such assumptions are known as Gaussian filters. They are also alternatively called quadrature filters, deterministic sample point filters etc.

To determine the first two moments, *viz.* mean and covariance, an integral in the form of $\int$ *nonlinear function* $\times$ *Gaussian pdf* must be evaluated. For an arbitrary nonlinear function, the integral becomes intractable and is generally solved approximately. An approximate solution involves representing the pdf with a few deterministic support points and the corresponding weights and manipulating the integral with a summation over those points. There are many methods to represent a Gaussian pdf with deterministic sample points and those will be discussed in detail in subsequent chapters. The general algorithm of the Gaussian filter is realized recursively with two steps, namely (i) time update and (ii) measurement update.

Time update: As stated above, we assume that the prior pdf of states is Gaussian i.e., $p(\mathcal{X}_k | \mathcal{Y}_{1:k-1}) \sim \mathcal{N}(\mathcal{X}_k; \hat{\mathcal{X}}_{k|k-1}, \Sigma_{k|k-1})$. Please note that, here $\hat{\mathcal{X}}_{k|k-1}$ means the estimated value (mean) of $\mathcal{X}_k$ given that the measurement up to the $(k-1)$ step has been received. The prior mean can be written as,

$$\begin{aligned}
\hat{\mathcal{X}}_{k|k-1} &= \mathbb{E}[\mathcal{X}_k | \mathcal{Y}_{1:k-1}] \\
&= \mathbb{E}[(\phi(\mathcal{X}_{k-1}, k) + \eta_k) | \mathcal{Y}_{1:k-1}] \\
&= \mathbb{E}[\phi(\mathcal{X}_{k-1}, k) | \mathcal{Y}_{1:k-1}],
\end{aligned}$$

or,

$$\hat{\mathcal{X}}_{k|k-1} = \int \phi(\mathcal{X}_{k-1}, k) p(\mathcal{X}_{k-1}|\mathcal{Y}_{1:k-1}) d\mathcal{X}_{k-1}$$

$$= \int \phi(\mathcal{X}_{k-1}, k) \mathcal{N}(\mathcal{X}_{k-1}; \hat{\mathcal{X}}_{k-1|k-1}, \Sigma_{k-1|k-1}) d\mathcal{X}_{k-1}. \tag{1.48}$$

Prior error covariance is given as,

$$\Sigma_{k|k-1} = \mathbb{E}[(\mathcal{X}_k - \hat{\mathcal{X}}_{k|k-1})(\mathcal{X}_k - \hat{\mathcal{X}}_{k|k-1})^T]$$

$$= \mathbb{E}[(\phi(\mathcal{X}_{k-1}, k) + \eta_k - \hat{\mathcal{X}}_{k|k-1})(\phi(\mathcal{X}_{k-1}, k) + \eta_k - \hat{\mathcal{X}}_{k|k-1})^T]$$

$$= \mathbb{E}[\phi(\mathcal{X}_{k-1}, k)\phi^T(\mathcal{X}_{k-1}, k) + \eta_k \eta_k^T + \hat{\mathcal{X}}_{k|k-1}\hat{\mathcal{X}}_{k|k-1}^T].$$

Please note that the expectation of all the cross terms is zero because the noise and the state are independent. The above equation is further written as

$$\Sigma_{k|k-1} = \int \phi(\mathcal{X}_{k-1}, k)\phi^T(\mathcal{X}_{k-1}, k)\mathcal{N}(\mathcal{X}_{k-1}; \hat{\mathcal{X}}_{k-1|k-1}, \Sigma_{k-1|k-1}) d\mathcal{X}_{k-1}$$

$$- \hat{\mathcal{X}}_{k|k-1}\hat{\mathcal{X}}_{k|k-1}^T + Q_k. \tag{1.49}$$

So, the approximated prior pdf becomes $p(\mathcal{X}_k|\mathcal{Y}_{1:k-1}) \sim \mathcal{N}(\mathcal{X}_k; \hat{\mathcal{X}}_{k|k-1}, \Sigma_{k|k-1})$.

Measurement update: As stated above, it is assumed that the pdf of measurements is Gaussian, i.e., $p(\mathcal{Y}_k|\mathcal{Y}_{1:k-1}) \sim \mathcal{N}(\mathcal{Y}_k; \hat{\mathcal{Y}}_{k|k-1}, \Sigma_{yy,k|k-1})$. The mean and covariance of measurement are given by

$$\hat{\mathcal{Y}}_{k|k-1} = \int \gamma(\mathcal{X}_k, k)\mathcal{N}(\mathcal{X}_k; \hat{\mathcal{X}}_{k|k-1}, \Sigma_{k|k-1}) d\mathcal{X}_k, \tag{1.50}$$

and

$$\Sigma_{yy,k|k-1} = \int \gamma(\mathcal{X}_k, k)\gamma^T(\mathcal{X}_k, k)\mathcal{N}(\mathcal{X}_k; \hat{\mathcal{X}}_{k|k-1}, \Sigma_{k|k-1}) d\mathcal{X}_k - \hat{\mathcal{Y}}_{k|k-1}\hat{\mathcal{Y}}_{k|k-1}^T + R_k. \tag{1.51}$$

Cross covariance of state and measurement is given by

$$\Sigma_{\mathcal{X}y,k|k-1} = \int \mathcal{X}_k \gamma^T(\mathcal{X}_k, k)\mathcal{N}(\mathcal{X}_k; \hat{\mathcal{X}}_{k|k-1}, \Sigma_{k|k-1}) d\mathcal{X}_k - \hat{\mathcal{X}}_{k|k-1}\hat{\mathcal{Y}}_{k|k-1}^T. \tag{1.52}$$

Throughout the book, the random variables on which the covariance is calculated are represented with a subscript and superscript of Σ, interchangeably. On the receipt of a new measurement $\mathcal{Y}_k$, the posterior density of state is given by

$$p(\mathcal{X}_k|\mathcal{Y}_{1:k}) = \mathcal{N}\left(\mathcal{X}_k; \hat{\mathcal{X}}_{k|k}, \Sigma_{k|k}\right), \tag{1.53}$$

where

$$\hat{\mathcal{X}}_{k|k} = \hat{\mathcal{X}}_{k|k-1} + K_k\left(\mathcal{Y}_k - \hat{\mathcal{Y}}_{k|k-1}\right), \tag{1.54}$$

$$\Sigma_{k|k} = \Sigma_{k|k-1} - K_k \Sigma_{\mathcal{Y}\mathcal{Y},k|k-1} K_k^T, \qquad (1.55)$$

$$K_k = \Sigma_{\mathcal{X}\mathcal{Y},k|k-1} \Sigma_{\mathcal{Y}\mathcal{Y},k|k-1}^{-1}. \qquad (1.56)$$

The Eq. (1.54) comes from general structure of the filter [20] and observer Eqs. (1.55) and (1.56) come from the Kalman filter which will be discussed in depth in the next chapter. With the assumption of the Gaussian prior and posterior pdf, we reach the expression of state estimate. Still the question of how we can evaluate the integral (1.48) to (1.52) remains unsolved.

Numerically, the Gaussian distribution is approximated with a few wisely chosen support points and weights associated with them. Approximately, those integrals are evaluated with a summation over those points. There are many methods available in the literature which describe how the deterministic points and weights should be chosen. The present book centers around those methods which will be described in depth in the upcoming chapters. Needless to say, accuracy of the estimation depends on the accuracy of the approximate evaluation of the above integrals.

1.8 Performance measure

While estimating the states of a dynamic system, a filter generally returns the posterior mean $\hat{\mathcal{X}}_{k+1|k+1}$ and its associated covariance $\Sigma_{k+1|k+1}$. As there are several different nonlinear filters available to the designer, a performance measure and comparison among the filters are necessary. The different methods that exist for evaluation of estimation algorithm are discussed in [103].

The consistency of the estimators can initially be established by checking the unbiasedness nature of the estimates. That is, the condition $\mathbb{E}(\mathcal{X}_{k+1} - \hat{\mathcal{X}}_{k+1|k+1}) = 0$ should hold. In addition to this, innovation, that is ($\mathcal{Y}_{k+1} - \hat{\mathcal{Y}}_{k+1|k}$), should also follow a zero mean white Gaussian process with measurement covariance as calculated by the filter. Throughout the book we evaluate the performance and compare them with other filters under the assumption that truth is known.

1.8.1 When truth is known

Here, two statistical properties of filters involving estimation error are described. Since for determining the estimation error, the truth values of the states are required this can only be achieved in simulation.

- Root mean square error (RMSE): RMSEs of states provide the standard deviation of the estimation error whereas MSEs provide the corresponding covariance. This is a widely used statistic to evaluate the filtering accuracy, as a lower value of RMSE directly implies high accuracy. RMSE at each time step over M ensembles calculated from Monte Carlo runs can be computed using the following expression

$$RMSE_k = \sqrt{\frac{1}{M} \sum_{i=1}^{M} (\mathcal{X}_{i,k} - \hat{\mathcal{X}}_{i,k|k})^2}. \qquad (1.57)$$

- Normalized state estimation error squared (NEES): This criteria is used to check the consistency of the filters. NEES is defined according to the relation $(\mathcal{X}_k - \hat{\mathcal{X}}_{k|k})^T \Sigma_{k|k}^{-1} (\mathcal{X}_k - \hat{\mathcal{X}}_{k|k})$ [100]. Under the hypothesis that the system is consistent and approximately linear Gaussian, this relation follows a chi-square distribution with n degrees of freedom, where n is the order of the system [13]. That is, its expected value will be the order of the system.

1.8.2 When truth is unknown

In the field, the truth value of the states is unknown. The only input given to the filter is noise corrupted measurements. In such a condition, an innovation sequence plays a key role in validating the filters. One statistical property used to measure the consistency of a filter is the normalized innovation squared (NIS), defined as $\nu_{k+1}^T \Sigma_{\mathcal{Y},k+1|k}^{-1} \nu_{k+1}$ [13]. Here, ν_{k+1} is the innovation and $\mathcal{Y}_{k+1} \sim \mathcal{N}(\hat{\mathcal{Y}}_{k+1|k}, \Sigma_{\mathcal{Y},k+1|k})$. Now NIS follows a chi-square distribution with p degrees of freedom, where p is the dimension of the measurement.

1.9 A few applications

The Kalman filter (and its extension to nonlinear system) is now being used in many fields of science and technology. In fact the application is so diversified that it is very difficult to summarize in a section. Here we provide a handful of applications.

1.9.1 Target tracking

Target tracking problems find their application in many civilian and military applications like air traffic control (ATC), spacecraft control, underwater target tracking etc. The first practical application of the Kalman filter was its implementation on a spacecraft mission to the moon and back, the Apollo project by NASA. The potential applicability of this algorithm was discovered by a NASA scientist, Stanley F. Schmidt, and this resulted in the formulation of the extended Kalman filter. The filter was used for the trajectory estimation and control of the spacecraft [59, 13]. Later, it also became an integral part of the onboard guidance module of the Apollo spacecraft. Since then, state estimation has become an integral part of most of the trajectory estimation and control modules designed.

1.9.2   Navigation

Nonlinear filters have become an integral part of navigation systems [13]. Their role is to find the vehicle position, attitude and velocity with the highest accuracy possible. In an inertial navigation system (INS), the ability of the Kalman filter to solve sensor fusion problems is exploited, where the global positioning system (GPS) and inertial sensor fusion is achieved. The inputs given by GPS and inertial sensors like the accelerometer and gyro are used for accurately obtaining the coordinates and orientation of the vehicle.

1.9.3   Process control

Use of the Kalman filter in process control such as the control of a chemical plant is widely studied [115, 61]. Recent studies have proposed estimating the tuning parameters of PID controllers online using the Kalman filter [131]. Here, the filter can receive input measurements from various sensors like pressure gauge, flow meters, temperature sensors, gas analyzers etc.

1.9.4   Weather prediction

There are applications where the future states of the system are predicted, but cannot be controlled as such. The Kalman filter plays a major role in predicting weather and natural disasters such as flood, tornado [72] etc. In the case of weather prediction, the scenario turns into a data assimilation problem where the state vector will have millions of states. The ensemble Kalman filter [43] has become a standard solution for weather prediction problems. Recently, this filter has been applied for predicting soil water monitoring and its contamination prediction [71].

1.9.5   Estimating state-of-charge (SoC)

Estimating the state-of-charge (SoC) of batteries online is a recent field of application of filtering [111]. It finds major application in electric vehicles where the battery management system should estimate the battery SoC, instantaneous power along with other parameters. These filtering problems are reported to be more challenging because of the fact that many parameter variations are not only time dependent but also user dependent.

The above mentioned applications and key sensors generally used for data fusion are tabulated in Table 1.1.

TABLE 1.1: Various applications of filtering

Applications	Dynamic system	Sensors
Target tracking	Spacecraft, submarine	RADAR, SONAR, imaging system
Navigation	Ship, missile, aircraft	Sextant, accelerometer, gyro, GPS
Process control	Chemical plant	pressure gauge, thermocouple, flow sensor, gas analyzer
Weather prediction	Atmosphere	Satellite, weather RADAR, sensors at weather station
Estimating SoC	Battery systems	Voltmeter

Apart from the above mentioned applications, state estimators find their role in finance [178], nonlinear structural dynamics [108], communication [134], biomedical engineering [18], power systems [139] to name a few. Many other industrial applications are discussed in [12] which is referred for further reading.

1.10 Prerequisites

There are prerequisites to understand the material presented in this book. These include (i) understanding of linear algebra and linear systems, (ii) Bayesian probability theory or at least multi-dimensional probability density and probability mass function and random process, (iii) state space analysis (both for continuous and discrete system). With the help of the above prerequisites, the first two chapters are easy to understand. Next the reader can either read all the chapters from 3 to 6, or any of them (based on necessity). In other words, Chapters 3-6 are not dependent upon each other. However, to read Chapters 7 to 9 understanding of the previous chapters is required.

In each chapter, we tried to provide an algorithm of the particular filter discussed there along with a sample MATLAB® code so that novice practitioners are not lost in the jungle of mathematical equations.

1.11 Organization of chapters

The book contains eight more chapters. Chapter 2 provides an optimal solution of the problem for a linear system and Gaussian noises. It also formulates the extended Kalman filter and its few important variants. Chapters 3 to 5 describe the unscented Kalman filter, cubature and quadrature based Kalman filters, the Gauss-Hermite filter and their variants respectively. Chapter 6 presents the Gaussian sum filter, where the prior and posterior pdfs are approximated with the weighted sum of several Gaussian pdfs. The next chapter considers a problem where measurements are randomly delayed. Such filters are finding more and more applications in networked control systems. Chapter 8 presents various filters for the continuous-discrete systems. Such systems naturally arise because process equations are in the continuous time domain as they are modeled from physical laws and the measurement equations are in the discrete time domain as they arise from the sensor measurement. Finally, in the last chapter two case studies, namely (i) bearing only underwater target tracking and (ii) tracking a ballistic target on reentry, have been considered. All the Gaussian filters are applied to them and results are compared.

Chapter 2

The Kalman filter and the extended Kalman filter

2.1   Linear Gaussian case (the Kalman filter) 27
    2.1.1   Kalman filter: a brief history 27
    2.1.2   Assumptions 28
    2.1.3   Derivation 29
    2.1.4   Properties: convergence and stability 31
    2.1.5   Numerical issues 32
    2.1.6   The information filter 33
    2.1.7   Consistency of state estimators 34
    2.1.8   Simulation example for the Kalman filter 35
    2.1.9   MATLAB®-based filtering exercises 37
2.2   The extended Kalman filter (EKF) 38
    2.2.1   Simulation example for the EKF 40
2.3   Important variants of the EKF 43
    2.3.1   The iterated EKF (IEKF) 43
    2.3.2   The second order EKF (SEKF) 45
    2.3.3   Divided difference Kalman filter (DDKF) 45
    2.3.4   MATLAB-based filtering exercises 49
2.4   Alternative approaches towards nonlinear filtering 49
2.5   Summary 50

2.1   Linear Gaussian case (the Kalman filter)

2.1.1   Kalman filter: a brief history

The linear Kalman filter (KF) is perhaps one of the most celebrated signal processing algorithms in practice. A complete solution to the filtering problem for discrete time state space systems with Gaussian noise was published in 1960 by Kalman [89]. A solution for continuous time case was presented by Kalman and Bucy in 1961 [90]. An extension of the Kalman filter for nonlinear systems (discussed later in section 2.2) was used for onboard guidance and navigation for the Apollo spacecraft. Since then, the Kalman filter has found applications across a wide variety of disciplines, including engineering, econometrics, quantitative finance, weather sciences and biomedical sciences.

The theory and applications of the KF and its extensions have been extensively discussed in the literature over the past 50 years; besides prominent foundational results in [89], see [90], [64], [35] and [11] for significant applications in different areas of science. [59] gives a historic overview of the Kalman filter and its applications in aerospace engineering over the first 50 years. [37] provides a succinct comparison of different filtering approaches, along with details on an exact nonlinear filtering approach under zero process noise.

The UKF and the associated deterministic sampling based filter is perhaps the newest family of filtering algorithms. Since [83] provided the first detailed description of the UKF algorithm, several modifications of the same have appeared (see [166], [87] and [120], among many others). Practical applications of the UKF in a variety of areas have been extensively discussed in the literature; see [48], [55] and [33] for example. The relationship between the EKF and the UKF is explored in [63], where it is shown that the EKF can also use deterministic sampling without explicit linearization. A separate strand of research within the research area of filtering based on deterministic sampling derives from computing the moments of nonlinear functions by computationally efficient means of explicit integration. These filters are called cubature Kalman filters (CKF) and cubature quadrature Kalman filters (CQKF); see [17], [5] and [16] for example. UKF, CKF and CQKF are discussed in more detail in subsequent chapters.

PF was first considered as a recursive Bayesian filter in [58]. This has been followed by major methodological advances (see, e.g., [8] and [40] for a summary of major developments) and some significant convergence results ([34]). Important applications of PF have been reported in [62], [28] and [112], among others.

In this chapter, we will provide an overview of the main results of the linear Kalman filter, and outline some extensions for nonlinear systems. Subsequent chapters will elaborate on the extensions which are the main focus of this book, *viz* filtering using assumed Gaussianity and deterministic sampling points.

2.1.2 Assumptions

Consider a linear state space system which can be expressed as

$$\mathcal{X}_{k+1} = A_k \mathcal{X}_k + b_k + \eta_k, \tag{2.1}$$

$$\mathcal{Y}_k = C_k \mathcal{X}_k + d_k + v_k, \tag{2.2}$$

where the matrices A_k, C_k and the vectors b_k, d_k are of compatible dimensions, are either constant or are known functions of time step k. Eq. (2.1) is commonly refrerred to as the transition equation, and Eq. (2.2) is the measurement equation. As in Chapter 1, v_k, η_k are zero mean, Gaussian and

$$\mathbb{E}(\eta_k \eta_m^T) = \mathbb{E}(\eta_k v_k^T) = \mathbb{E}(v_k v_m^T) = 0$$

hold. The constant *offset* vectors b, d occur in many econometric and financial applications, and they are also relevant in engineering applications when we consider linearization of nonlinear systems. The covariance matrices of noise terms are given by

$$\mathbb{E}(\eta_k \eta_k^T) = Q_k, \ \mathbb{E}(v_k v_k^T) = R_k.$$

As in the previous chapter, we will omit the 'control input' term $\mathcal{U}_k$ in the first equation, which can be designed as a linear function of the state $\mathcal{X}_k$. We also assume that, for random variable $\mathcal{X}_0$, $\mathbb{E}(\mathcal{X}_0) = \hat{\mathcal{X}}_{0|0}$ and $\mathbb{E}(\mathcal{X}_0 - \hat{\mathcal{X}}_0)(\mathcal{X}_0 - \hat{\mathcal{X}}_0)^T = \Sigma_{0|0}$ are known.

2.1.3 Derivation

Our interest is to estimate the predictive and the posterior distributions i.e., $p(\mathcal{X}_k|\mathcal{Y}_{k-1})$ and $p(\mathcal{X}_k|\mathcal{Y}_k)$, at each time t_k.

Since all the variables are Gaussian and since Gaussian distributions are completely characterized by the mean vector and the covariance matrix, this problem amounts to finding four quantities $\mathbb{E}(\mathcal{X}_k|\mathcal{Y}_{k-1})$, $\mathbb{E}(\mathcal{X}_k\mathcal{X}_k^T|\mathcal{Y}_{k-1})$, $\mathbb{E}(\mathcal{X}_k|\mathcal{Y}_k)$, $\mathbb{E}(\mathcal{X}_k\mathcal{X}_k^T|\mathcal{Y}_k)$.

The *Kalman filter* is a tool to do this *recursively*, by combining the model prediction at time t_{k-1} with the new measurement $\mathcal{Y}_k$ at time t_k, to obtain the estimates at time t_k. We now proceed to derive the necessary recursive equations as follows.

- Define

$$\hat{\mathcal{X}}_{k|k-1} = \mathbb{E}(\mathcal{X}_k|\mathcal{Y}_{k-1}), \ \Sigma_{k|k-1}^{\mathcal{X}\mathcal{X}} = \mathbb{E}(\mathcal{X}_k - \hat{\mathcal{X}}_{k|k-1})(\mathcal{X}_k - \hat{\mathcal{X}}_{k|k-1})^T,$$

$$\hat{\mathcal{X}}_{k|k} = \mathbb{E}(\mathcal{X}_k|\mathcal{Y}_k), \ \Sigma_{k|k}^{\mathcal{X}\mathcal{X}} = \mathbb{E}(\mathcal{X}_k - \hat{\mathcal{X}}_{k|k})(\mathcal{X}_k - \hat{\mathcal{X}}_{k|k})^T.$$

- Assume that $\hat{\mathcal{X}}_{k-1|k-1}, \Sigma_{k-1|k-1}^{\mathcal{X}\mathcal{X}}$ are already available.

- Using the known transition equation, we can carry a **time update** of the mean and the covariance of the unobservable state:

$$\hat{\mathcal{X}}_{k|k-1} = A\hat{\mathcal{X}}_{k-1|k-1} + b, \text{ and} \tag{2.3}$$

$$\Sigma_{k|k-1}^{\mathcal{X}\mathcal{X}} = A\Sigma_{k-1|k-1}^{\mathcal{X}\mathcal{X}}A^T + Q_k. \tag{2.4}$$

This covariance matrix update formula was mentioned earlier in Chapter 1. Note that η_k is independent of $\mathcal{X}_{k-1}$.

- When $\mathcal{Y}_k$ becomes available, the *innovation* (or new information) in $\mathcal{Y}_k$ is

$$\mathcal{Y}_k - \underbrace{(C\hat{\mathcal{X}}_{k|k-1} + d)}_{\hat{\mathcal{Y}}_{k|k-1}} = C(\mathcal{X}_k - \hat{\mathcal{X}}_{k|k-1}) + v_k.$$

Note that the covariance of innovations is given by

$$\mathbb{E}(\mathcal{Y}_k - \hat{\mathcal{Y}}_{k|k-1})(\mathcal{Y}_k - \hat{\mathcal{Y}}_{k|k-1})^T = C\Sigma^{\mathcal{X}\mathcal{X}}_{k|k-1}C^T + R_k. \qquad (2.5)$$

- One can now look at the *joint* Gaussian density of $\mathcal{X}_k$ and $\mathcal{Y}_k$:

$$\begin{bmatrix} \mathcal{X}_k \\ \mathcal{Y}_k \end{bmatrix} \sim \mathcal{N}\left(\begin{bmatrix} \hat{\mathcal{X}}_{k|k-1} \\ \hat{\mathcal{Y}}_{k|k-1} \end{bmatrix}, \begin{bmatrix} \Sigma^{\mathcal{X}\mathcal{X}}_{k|k-1} & \Sigma^{\mathcal{X}\mathcal{X}}_{k|k-1}C^T \\ C\Sigma^{\mathcal{X}\mathcal{X}}_{k|k-1} & C\Sigma^{\mathcal{X}\mathcal{X}}_{k|k-1}C^T + R_k \end{bmatrix} \right), \qquad (2.6)$$

where we use

$$\begin{aligned} \mathbb{E}(\mathcal{X}_k - \hat{\mathcal{X}}_{k|k-1})&(\mathcal{Y}_k - \hat{\mathcal{Y}}_{k|k-1})^T \\ &= \mathbb{E}(\mathcal{X}_k - \hat{\mathcal{X}}_{k|k-1})(C(\mathcal{X}_k - \hat{\mathcal{X}}_{k|k-1}) + v_{k-1})^T \\ &= \Sigma^{\mathcal{X}\mathcal{X}}_{k|k-1}C^T, \end{aligned}$$

since v_k is uncorrelated with $\mathcal{X}_k$.

- We now use the following standard result on Gaussian random variables from statistics (see, e.g., [60] for a proof): Let $x \sim \mathcal{N}(\hat{x}, \Sigma^{xx})$, $y \sim \mathcal{N}(\hat{y}, \Sigma^{yy})$ be jointly Gaussian with a cross-covariance matrix Σ^{xy}. Then the first two conditional moments of x are given by

$$\mathbb{E}(x|y) = \hat{x} + \Sigma^{xy}(\Sigma^{yy})^{-1}(y - \mathbb{E}y), \quad \text{and}$$
$$\mathbb{E}(x - \mathbb{E}(x|y))(x - \mathbb{E}(x|y))^T = \Sigma^{xx} - \Sigma^{xy}(\Sigma^{yy})^{-1}\Sigma^{yx}. \qquad (2.7)$$

In the present case, this leads to

$$\hat{\mathcal{X}}_{k|k} = \hat{\mathcal{X}}_{k|k-1} + K_k(\mathcal{Y}_k - \hat{\mathcal{Y}}_{k|k-1}), \quad \text{and} \qquad (2.8)$$
$$\Sigma^{\mathcal{X}\mathcal{X}}_{k|k} = \Sigma^{\mathcal{X}\mathcal{X}}_{k|k-1} - \Sigma^{\mathcal{X}\mathcal{X}}_{k|k-1}C^T \left(C\Sigma^{\mathcal{X}\mathcal{X}}_{k|k-1}C^T + R_k\right)^{-1} C\Sigma^{\mathcal{X}\mathcal{X}}_{k|k-1}, \qquad (2.9)$$

where

$$K_k := \Sigma^{\mathcal{X}\mathcal{X}}_{k|k-1}C^T \left(C\Sigma^{\mathcal{X}\mathcal{X}}_{k|k-1}C^T + R_k\right)^{-1} \qquad (2.10)$$

is called the **Kalman gain**. Eqs. (2.3) - (2.4) and (2.8) - (2.10) completely define the Kalman filter recursions, with the last three equations defining the **measurement update** step of the filter.

- Alternatively, suppose that z_k is any estimate of x_k which is of the form

$$z_k = \hat{\mathcal{X}}_{k|k-1} + \Phi_k(\mathcal{Y}_k - \hat{\mathcal{Y}}_{k|k-1}).$$

We can then choose to proceed with the constant matrix gain Φ_k such that z_k is the best estimate from a minimum variance point of view. By carrying out manipulations similar to those in Section 1.7.1, one can get an expression for the covariance matrix Σ_k^{zz} in terms of Φ_k in this case:

$$
\Sigma_{k|k}^{zz} = \mathbb{E}(\mathcal{X}_k - z_k)(\mathcal{X}_k - z_k)^T
$$

$$
= \mathbb{E}[(\mathcal{X}_k - \hat{\mathcal{X}}_{k|k-1}) - \Phi_k(\mathcal{Y}_k - \hat{\mathcal{Y}}_{k|k-1})][(\mathcal{X}_k - \hat{\mathcal{X}}_{k|k-1}) \tag{2.11}
$$

$$
- \Phi_k(\mathcal{Y}_k - \hat{\mathcal{Y}}_{k|k-1})]^T
$$

$$
= \Sigma_{k|k-1}^{\mathcal{X}\mathcal{X}} + \Phi_k(C\Sigma_{k|k-1}^{\mathcal{X}\mathcal{X}}C^T + R_k)\Phi_k^T - \Sigma_{k|k-1}^{\mathcal{X}\mathcal{X}}C^T\Phi_k^T - \Phi_k C\Sigma_{k|k-1}^{\mathcal{X}\mathcal{X}}. \tag{2.12}
$$

By using standard results from matrix calculus, it can be shown that the trace of $\Sigma_{k|k}^{zz}$ is minimized by choosing $\Phi_k = K_k$. In other words, Kalman gain is a minimum variance estimator which is linear in $\mathcal{Y}_k$. This property holds even if the noise variables are not Gaussian.

- We need to initialize the filter at time $k = 0$. This is done by assuming that $\mathcal{X}_0$ is a Gaussian random variable with a known mean $\hat{\mathcal{X}}_{0|0}$ and a known variance $\Sigma_{0|0}^{\mathcal{X}\mathcal{X}}$. These quantities may be known or inferred from the physics of the problem. It is customary to take $\Sigma_{0|0}^{\mathcal{X}\mathcal{X}}$ as a large constant multiplied by identity.

2.1.4 Properties: convergence and stability

We will now discuss some of the key properties of the recursive filtering algorithm described in the previous section.

- For constant Q, R, the propagation of the conditional covariance matrix can be written as

$$
\Sigma_{k|k}^{\mathcal{X}\mathcal{X}} = A\Sigma_{k-1|k-1}^{\mathcal{X}\mathcal{X}}A^T + Q
$$

$$
- A\Sigma_{k-1|k-1}^{\mathcal{X}\mathcal{X}}C^T(C\Sigma_{k-1|k-1}^{\mathcal{X}\mathcal{X}}C^T + R)^{-1}C\Sigma_{k-1|k-1}^{\mathcal{X}\mathcal{X}}A^T. \tag{2.13}
$$

Interestingly, note that this depends solely on the system properties and not on the measurements. The sequence $\Sigma_{k|k}^{\mathcal{X}\mathcal{X}}$ can thus be pre-computed up to any arbitrary value of k, before receiving any measurement. Further, we can take limit as $k \to \infty$ to get

$$
\Sigma_\infty^{\mathcal{X}\mathcal{X}} = A\Sigma_\infty^{\mathcal{X}\mathcal{X}}A^T + Q
$$

$$
- A\Sigma_\infty^{\mathcal{X}\mathcal{X}}C^T(C\Sigma_\infty^{\mathcal{X}\mathcal{X}}C^T + R)^{-1}C\Sigma_\infty^{\mathcal{X}\mathcal{X}}A^T. \tag{2.14}
$$

Eq. (2.14) is called the discrete algebraic Riccati equation.

- The state space system of the predictive operation of our estimator can also be written as

$$\hat{\mathcal{X}}_{k|k-1} = A \overbrace{\left(\hat{\mathcal{X}}_{k-1|k-2} + K_{k-1}\mathcal{Y}_{k-1} - K_{k-1}C\hat{\mathcal{X}}_{k-1|k-2} \right)}^{\hat{\mathcal{X}}_{k-1|k-1}} + b$$

$$= (A - K_{k-1}C)\hat{\mathcal{X}}_{k-1|k-2} + b + K_{k-1}\mathcal{Y}_{k-1}. \qquad (2.15)$$

To be able to use the estimator, we need the matrix $(A - K_{k-1}C)$ to be stable (at least asymptotically); i.e., it should have all its eigenvalues inside the unit circle. This follows from the earlier discussion on stability in Chapter 1. This depends on the asymptotic Kalman gain K_∞, which in turn depends on $\Sigma_\infty^{\mathcal{X}\mathcal{X}}$. To state the result related to stability of $(A - K_\infty C)$, we need two definitions:

Definition 1 *A pair of matrices* $C \in \mathbb{R}^{m \times n}, A \in \mathbb{R}^{n \times n}$ *is observable if* $\begin{bmatrix} C & CA & \cdots & CA^{n-1} \end{bmatrix}^T$ *has full column rank.*

Definition 2 *A pair of matrices* $A \in \mathbb{R}^{n \times n}, Q \in \mathbb{R}^{n \times m}$ *is controllable if* $\begin{bmatrix} Q & AQ & \cdots & A^{n-1}Q \end{bmatrix}$ *has full row rank.*

Lemma 1 *If* (C, A) *is observable and* (A, Q) *is controllable, then there is a solution for the DARE such that* $A - K_\infty C$ *is **guaranteed** to be stable; i.e., it has all its eigenvalues inside the unit circle in the complex plane.*

This result may be seen as a key to the success of the Kalman filter in engineering applications; it yields a **time-invariant** implementation which is guaranteed to be stable.

2.1.5   Numerical issues

The covariance update Eq. (2.9) requires taking a difference between two positive definite matrices, which may lead to a loss of positive definiteness. This can be avoided (at the cost of extra computation) by reorganizing (2.9) as

$$\Sigma_{k|k}^{\mathcal{X}\mathcal{X}} = (I - K_kC)\Sigma_{k|k}^{\mathcal{X}\mathcal{X}}(I - K_kC)^T + K_kR_kK_k^T. \qquad (2.16)$$

This covariance update form is called the **Joseph form**. The equivalence of the Joseph form to the covariance update equation for a generic gain (2.12) should be obvious.

Another approach to improve the numerical stability is to use **square root filters**, which propagate a square root of the predicted covariance matrix $\Sigma_{k|k-1}^{\mathcal{X}\mathcal{X}}$ from one time step to the next, instead of propagating the matrix itself. We outline one common form of square root filter here, with specialized forms to follow in subsequent chapters. If $\Sigma_{k|k-1}^{\mathcal{X}\mathcal{X}} = S_k S_k^T$ for a real lower triangular matrix S_k, the square root recursion can be encapsulated in the following equation (see, e.g., [2] and [172], whose notation is followed here):

$$\underbrace{\begin{bmatrix} R_k^{\frac{1}{2}} & C_k S_k & 0 \\ 0 & A_k S_k & B_k Q_k^{\frac{1}{2}} \end{bmatrix}}_{\text{pre-array}} U = \underbrace{\begin{bmatrix} \tilde{R}_{k+1}^{\frac{1}{2}} & 0 & 0 \\ G_{k+1} & S_{k+1} & 0 \end{bmatrix}}_{\text{post-array}}, \qquad (2.17)$$

where U is an orthogonal transformation that triangularizes the pre-array. One can construct U using the Householder transformation [70]. Given the pre-array, one can find the orthogonal transformation U using the Householder transformation [70] and hence find the lower triangular matrices $\tilde{R}_{k+1}$, S_{k+1} and the matrix G_{k+1} in the post-array. The filtering equation is then given by

$$\hat{\mathcal{X}}_{k+1|k} = A\hat{\mathcal{X}}_{k|k-1} + G_{k+1}\tilde{R}_{k+1}^{-\frac{1}{2}}(\mathcal{Y}_k - \hat{\mathcal{Y}}_{k|k-1}), \qquad (2.18)$$

with

$$\hat{\mathcal{Y}}_{k|k-1} = C\hat{\mathcal{X}}_{k|k-1} + d, \qquad \text{as before.}$$

Updating the square root of the matrix is numerically efficient and also avoids the problem with the covariance matrix losing positive semi-definiteness. Alternative methods of square root filtering are discussed in [172], along with a detailed analysis of computational complexity and errors in covariance computation.

   If the loss of symmetry of the covariance matrix due to numerical errors is an issue, a simple way of avoiding it is to compute full $\Sigma_{k|k}^{\mathcal{X}\mathcal{X}}$ as in Eq. (2.9) and then reset its value as

$$\Sigma_{k|k}^{\mathcal{X}\mathcal{X}} := 1/2(\Sigma_{k|k}^{\mathcal{X}\mathcal{X}} + (\Sigma_{k|k}^{\mathcal{X}\mathcal{X}})^T). \qquad (2.19)$$

2.1.6   The information filter

   Instead of recursively updating the covariance matrix and the state estimate for a linear system, the information filter implementation of the Kalman filter recursively updates the inverse of the covariance matrix, $\Phi \triangleq (\Sigma^{\mathcal{X}\mathcal{X}})^{-1}$ and an estimate of a transformed state given by $\mathcal{X}^I \triangleq \Sigma^{\mathcal{X}\mathcal{X}}\mathcal{X}$. We provide the set of equations for updating $(\hat{\mathcal{X}}^I{}_{k-1|k-1}, \Phi_{k-1|k-1}^{xx}) \mapsto (\hat{\mathcal{X}}^I{}_{k|k}, \Phi_{k|k}^{xx})$ at time step k without proof below.

$$\Psi_k \triangleq (A^{-1})^T \Phi_{k-1|k-1} A^{-1},$$

$$\Phi_{k|k-1} = (I + \Psi_k Q_k)^{-1}\Psi_k,$$

$$\hat{\mathcal{X}}^I{}_{k|k-1} = (I + \Psi_k Q_k)^{-1}(A^T)^{-1}\hat{\mathcal{X}}^I{}_{k-1|k-1},$$

$$\hat{\mathcal{X}}^I{}_{k|k} = \hat{\mathcal{X}}^I{}_{k|k-1} + C^T R_k^{-1}\mathcal{Y}_k,$$

$$\Phi_{k|k} = \Phi_{k|k-1} + C^T R_k^{-1}C.$$

Here, we have assumed $b = 0$ for simplicity. The advantage of the information filter is the ease of initialisation; if there is no prior information available about the state at $k = 0$, one can take $\Phi_0 = 0$. It is not possible to set $\Sigma^{\mathcal{X}\mathcal{X}} = \infty$ in the Kalman filter. The disadvantages are the loss of intuition (since the covariance matrix itself is not updated) and the additional requirement that A should be invertible.

2.1.7   Consistency of state estimators

In practice, as the latent state is unavailable (by definition), we don't have the state estimation error to make a judgement about how well the filter is performing. It is possible for a Kalman filter to *diverge*, due to errors in modelling or programming. One can use the available information, i.e., the innovations sequence and the sequence of covariance matrices, to make a judgement about whether the filter is consistent with the assumptions on the underlying model. The model assumptions imply that $p(\mathcal{X}_k | \mathcal{Y}_{1:k})$ has a normal distribution with a mean $\hat{\mathcal{X}}_{k|k}$ and a covariance $\Sigma_{k|k}^{\mathcal{X}\mathcal{X}}$. Further, the assumptions imply that the innovations $\epsilon_k := \mathcal{Y}_k - \hat{\mathcal{Y}}_{k|k-1}$ have mean zero, covariance given by (2.5) and are independent from one another. Given a single test run over a finite number of samples K, one can carry out the following statistical tests to check for possible inconsistencies in the filter implementation:

- First compute the following test statistic:

$$\bar{\epsilon} := \frac{1}{N} \sum_{k=1}^{N} \epsilon_k^T (\Sigma_k^{yy})^{-1} \epsilon_k,$$

  where $\Sigma_{k|k-1}^{yy} = C\Sigma_{k|k-1}^{\mathcal{X}\mathcal{X}} C^T + R_k$ is the covariance of innovations. Under the model assumptions about linearity, Gaussian disturbances and ergodicity of disturbances (which allows us to replace ensemble averages with time averages), $N\bar{\epsilon}$ has a chi-squared distribution with mN degrees of freedom, where m is the dimension of the measurement vector $\mathcal{Y}_k$. One can then evaluate whether the observed value of $\bar{\epsilon}$ falls within a prescribed confidence interval (say, 95%) for this distribution.

- Also compute the test statistic for correlation across the time steps:

$$\bar{\rho}_{j,i} = \frac{\sum_{k=1}^{N-j} \epsilon_{k,i} \epsilon_{k+j,i}}{\sqrt{\sum_{k=1}^{N-j} \epsilon_{k,i}^2 \epsilon_{k+j,i}^2}},$$

  where $\epsilon_{k,i}$ is the ith element of the innovations vector ϵ_k. Under model assumptions, $\bar{\rho}_{j,i}$ is normally distributed with mean zero and covariance $1/N$ for each j.

2.1.8 Simulation example for the Kalman filter

Consider a multidimensional state space system:

$$
\underbrace{\begin{bmatrix} x_k \\ x_k^v \\ y_k \\ y_k^v \end{bmatrix}}_{X_k} = \begin{bmatrix} 1 & \frac{\sin \Omega T}{\Omega} & 0 & -\frac{(1-\cos(\Omega T))}{\Omega} \\ 0 & \cos(\Omega T) & 0 & -\sin(\Omega T) \\ 0 & -\frac{(1-\cos(\Omega T))}{\Omega} & 1 & -\frac{\sin(\Omega T)}{\Omega} \\ 0 & \sin(\Omega T) & 0 & \cos(\Omega T) \end{bmatrix} \begin{bmatrix} x_{k-1} \\ x_{k-1}^v \\ y_{k-1} \\ y_{k-1}^v \end{bmatrix} + \eta_{k-1}.
$$

The measurement equation is

$$
\mathcal{Y}_k = \begin{bmatrix} x_k \\ y_k \end{bmatrix} + v_k.
$$

Here, v_k, η_k are i.i.d. zero mean random variables with $\mathbb{E}(\eta_k \eta_k^T) = Q$, $\mathbb{E}(v_k v_k^T) = R$. The parameters of the system are as follows:

$$
\Omega = \pi/4, \ T = 0.5, \ q = 0.1,
$$

$$
Q = q \begin{bmatrix} T^3/3 & T^2/2 & 0 & 0 \\ T^2/2 & T & 0 & 0 \\ 0 & 0 & T^3/3 & T^2/2 \\ 0 & 0 & T^2/2 & T \end{bmatrix},
$$

$$
R = \begin{bmatrix} 120^2 & 0 \\ 0 & 70 \end{bmatrix}.
$$

The initial state is assumed to be

$$
\mathcal{X}_0 = \begin{bmatrix} 1000 & 30 & 1000 & 0 \end{bmatrix}^T,
$$

with the initial covariance

$$
\Sigma_0^{XX} = \begin{bmatrix} 200 & 0 & 0 & 0 \\ 0 & 20 & 0 & 0 \\ 0 & 0 & 200 & 0 \\ 0 & 0 & 0 & 20 \end{bmatrix}.
$$

The code provided performs the following tasks.

(i) Write a state space simulation for the above system, which simulates the true measurements and the true state for 200 time-steps.

(ii) Implement the Kalman filter for the system to estimate the unobservable states from measurements.

(iii) Repeat the above simulation 100 times. to find the root mean squared error *at each time step* for x_k^v and y_k^v. Plot these errors against time.

```
1 % Linear KF for simplified constant turn model
2 clear;
3 N=200; % number of simulations
4 M=100; %number of time steps in each simulation.
5 %--- Define parameter values
6 Omega= pi/4; T=0.5;   q=0.1;
7 Q= q*[T^3/3 T^2/2 0 0;T^2/2 T 0 0; 0 0 T^3/3 T^2/2;
8        0 0 T^2/2 T];
9 Qs = sqrtm(Q);
10 Rs=[120 0;0 abs(sqrt(70))];
11 R=Rs*Rs';
12 x0=[1000; 30; 1000;0]; P0=diag([200;20;200;20]);
13 A=[1 sin(Omega*T)/Omega    0  -(1-cos(Omega*T))/Omega;
14    0 cos(Omega*T)          0  -sin(Omega*T);
15    0 (1-cos(Omega*T))/Omega 1   sin(Omega*T)/Omega;
16    0   sin(Omega*T)         0   cos(Omega*T)];
17    C= [ 1 0 0 0 ; 0 1 0 0];
18 x=[x0 zeros(4,N-1)];
19
20
21 telapsed=0;
22 for j=1:M
23 %--- Simulate the true system and run the KF M times
24 xhat=[x0+sqrtm(P0)*randn(size(x0)) zeros(4,N-1)];
25 xhat_minus=xhat;
26 P_new=P0;
27 P_old=P0;
28 y=zeros(2,N);
29 se1=zeros(M,N); se2=se1;
30 y(:,1)=C*x0+R*randn(2,1);
31 for i=1:N-1 %one sample path
32     x(:,i+1)=A*x(:,i)+Qs*randn(4,1);
33     y(:,i+1)=C*x(:,i+1)+Rs*randn(2,1);
34 end
35
36 %--- Kalman filter implementation ---
37 xhat=[x0+sqrtm(P0)*randn(size(x0)) zeros(4,N-1)];
38
39 tstart=tic; %---start measuring time.---
40 for i=1:N-1
41     xhat_minus(:,i+1)= A*xhat(:,i);
42     P_old = A*P_new*A' + Q;
43     K=P_old*C'*(C*P_old*C'+R)^(-1);
44     Innovations=y(:,i+1)-C*xhat_minus(:,i+1);
45     xhat(:,i+1)=xhat_minus(:,i+1)+K*innovations;
46     P_C_K = P_old*C'*K';
47     P_new=P_old + K*C*P_C_K +K*R*K'-P_C_K-P_C_K';
48 end
49 se1(j,1:N)=(xhat(2,:)-x(2,:)).^2;
50 se2(j,1:N)=(xhat(4,:)-x(4,:)).^2;
51 telapsed=telapsed+toc(tstart);
52 end %---end for the simulation counter---
```

```
53
54  rmse1=zeros(N,1);rmse2=rmse1;
55 %— average root mean squared errors at each
56            % time step, over M simulations
57  for i=1:N
58 rmse1(i) = sqrt(sum(se1(1:M,i))/M);
59 rmse2(i) = sqrt(sum(se2(1:M,i))/M);
60 end
61
62 average_time = telapsed/M;
63 %— this is average time per Kalman filter run
64 close all;
65
66 figure;plot(1:N, rmse1, 1:N,rmse2, '— ');
67 legend( 'RMSE for x^v_k ', 'RMSE for y^v_k ')
```

Listing 2.1: KF for simplified constant turn model.

A sample MATLAB code for KF is provided in Listing 2.1. The resulting plots of average errors are shown in Figure 2.1.

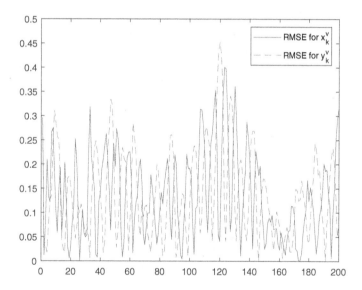

FIGURE 2.1: Root mean squared errors at each time step.

2.1.9 MATLAB®-based filtering exercises

(i) Repeat the numerical experiments in Section 2.1.8 with different values of parameters; e.g., see how changing the size of noise affects the errors.

(ii) Implement the consistency tests given in Section 2.1.7 for the system in Section 2.1.8 for a single simulation run.

(iii) If more than one test run is possible, one can carry out more sophisticated consistency tests on the Kalman filter, as mentioned in Section
2.1.7. Specifically, let ϵ_k^i and $\Sigma_k^{yy,i}$ be the innovation and its covariance
respectively, at k^{th} time-step and at i^{th} test run. Over N test runs, one
can compute the test statistic

$$\bar{\epsilon}_k := \frac{1}{N} \sum_{i=1}^{N} (\epsilon_k^i)^T (\Sigma_k^{yy,i})^{-1} \epsilon_k^i,$$

at *each* time step k, as an ensemble average (rather than as a time
average). $\bar{\epsilon}_k$ is chi-squared distributed with mN degrees of freedom. Use
100 simulation runs, each with 200 time steps, to plot 95% confidence
intervals for $\bar{\epsilon}_k$.

2.2   The extended Kalman filter (EKF)

The previous section described a scenario where the true physical system
can adequately be described in a linear state space framework, across its entire
operating range. For many physical systems, this is an unrealistic assumption.
A large class of physical systems can be represented by a nonlinear state space
representation of the form

$$\mathcal{X}_{k+1} = \phi_k(\mathcal{X}_k) + \eta_{k-1},$$
$$\mathcal{Y}_k = \gamma_k(\mathcal{X}_k) + v_k,$$

where ϕ_k, γ_k are known functions, $\{\eta_k\}, \{v_k\}$ are mutually independent ran-
dom processes with $\mathbb{E}(\eta_k) = \mathbb{E}(v_k) = 0$, $\mathbb{E}(\eta_k \eta_k^T) = Q_k$, $\mathbb{E}(v_k v_k^T) = R_k$ and
$\mathbb{E}(\eta_k \eta_j^T) = \mathbb{E}(v_k v_j^T) = 0$ if $k \neq j$. As before, $\mathcal{Y}_k$ are measurements from which
the information about $\mathcal{X}_k$ is to be inferred. If the functions ϕ_k, γ_k describing
the dynamics of the real system are continuously differentiable in $\mathcal{X}_k$, it might
still be possible to linearize the dynamics *locally* around the current state at
each time step using a first order Taylor series approximation:

$$\phi_k(\mathcal{X}_k) \approx \phi_k(\hat{\mathcal{X}}_{k|k}) + \nabla_{\phi_k}(\mathcal{X}_k - \hat{\mathcal{X}}_{k|k}), \qquad (2.20)$$

$$\gamma_k(\mathcal{X}_k) \approx \gamma_k(\hat{\mathcal{X}}_{k|k}) + \nabla_{\gamma_k}(\mathcal{X}_k - \hat{\mathcal{X}}_{k|k}), \qquad (2.21)$$

and then apply the Kalman filter to the linearized system. Here, ∇_{ϕ_k} and ∇_{γ_k}
are Jacobians computed at the current state estimate:

$$[\nabla_{\phi_k}]_{ij} = \left[\frac{\partial \phi_{ik}}{\partial \mathcal{X}_j} \right]_{\mathcal{X} = \hat{\mathcal{X}}_{k|k}}, \quad [\nabla_{\gamma_k}]_{ij} = \left[\frac{\partial \gamma_{i,k}}{\partial \mathcal{X}_j} \right]_{\mathcal{X} = \hat{\mathcal{X}}_{k|k}}, \qquad (2.22)$$

where $\phi_{i,k}, \gamma_{i,k}$ represent i^{th} elements of the corresponding vector valued functions, at k^{th} time-step. This technique is called the extended Kalman filter

---

Algorithm 4 Extended Kalman filter

$$[\hat{\mathcal{X}}_{k|k}, \Sigma_{k|k}] = \text{EKF}[\mathcal{Y}_k]$$

- Initialize the filter with $\hat{\mathcal{X}}_{0|0}$ and $\Sigma_{0|0}$.

- For $k = 1 : N$

    $$\hat{\mathcal{X}}_k^i = \phi_k(\hat{\mathcal{X}}_{k-1|k-1}),$$

    evaluate the Jacobians $\nabla_{\phi_{k-1}}, \nabla_{\gamma_{k-1}},$

    $$\Sigma_{k|k-1}^{\mathcal{X}\mathcal{X}} = \nabla_{\phi_{k-1}} \Sigma_{k-1|k-1}^{\mathcal{X}\mathcal{X}} \nabla_{\phi_{k-1}}^T + Q_k, \tag{2.23}$$

    $$K_k = \Sigma_{k|k-1}^{\mathcal{X}\mathcal{X}} \nabla_{\gamma_{k-1}}^T (\nabla_{\gamma_{k-1}} \Sigma_{k|k-1}^{\mathcal{X}\mathcal{X}} \nabla_{\gamma_{k-1}}^T + R_k)^{-1}, \tag{2.24}$$

    $$\hat{\mathcal{X}}_{k|k} = \hat{\mathcal{X}}_{k|k-1} + K_k(\mathcal{Y}_k - \gamma_k(\hat{\mathcal{X}}_{k|k-1}, k)), \tag{2.25}$$

    $$\Sigma_{k|k}^{\mathcal{X}\mathcal{X}} = \Sigma_{k|k-1}^{\mathcal{X}\mathcal{X}} + K_k \nabla_{\gamma_{k-1}} \Sigma_{k|k-1}^{\mathcal{X}\mathcal{X}} \nabla_{\gamma_{k-1}}^T K_k^T + K_k R_k K_k^T$$
    $$- \Sigma_{k|k-1}^{\mathcal{X}\mathcal{X}} \nabla_{\gamma_{k-1}}^T K_k^T - K_k \nabla_{\gamma_{k-1}} \Sigma_{k|k-1}. \tag{2.26}$$

- End For

(EKF). We describe the steps for the EKF below. The following points can be noted.

- It can be easily seen that the EKF is a time-varying implementation of the Kalman filter where the Jacobians ∇_{ϕ_k} and ∇_{γ_k} play the role of the transition matrix A and the measurement matrix C, respectively. The computation of Jacobians at each time step makes the filter time-varying; further, unlike the linear KF, it is impossible to pre-compute the sequence of covariance matrices *a priori* for the EKF. In some applications, it might still be possible to compute this sequence for a *nominal* trajectory.

- The convergence and stability properties of the KF are not preserved in the EKF. It is worth exploring this point further in light of the vector valued Taylor series expansion used in the EKF. Using Taylor's theorem, we can write

$$\phi_k(\mathcal{X}_k) = \phi_k(\hat{\mathcal{X}}_{k|k}) + \nabla_{\phi_k}(\mathcal{X}_k - \hat{\mathcal{X}}_{k|k}) + \tilde{\phi}_k(\mathcal{X}_k)(\mathcal{X}_k - \hat{\mathcal{X}}_{k|k}) + \eta_k, \tag{2.27}$$

$$\gamma_k(\mathcal{X}_k) = \gamma_k(\hat{\mathcal{X}}_{k|k}) + \nabla_{\gamma_k}(\mathcal{X}_k - \hat{\mathcal{X}}_{k|k}) + \tilde{\gamma}_k(\mathcal{X}_k)(\mathcal{X}_k - \hat{\mathcal{X}}_{k|k}) + v_k, \tag{2.28}$$

where the functions $\tilde{\phi}_k, \tilde{\gamma}_k$ are such that

$$\lim_{\mathcal{X}_k \to \hat{\mathcal{X}}_{k|k}} \tilde{\phi}_k(\mathcal{X}_k) = \lim_{\mathcal{X}_k \to \hat{\mathcal{X}}_{k|k}} \tilde{\gamma}_k(\mathcal{X}_k) = 0.$$

However, there is no guarantee that $\mathbb{E}(\mathcal{X}_k - \hat{\mathcal{X}}_{k|k}) = 0$, and the covariance matrix of $\mathcal{X}_k - \hat{\mathcal{X}}_{k|k}$ may not be given by $\Sigma_{k|k}^{\mathcal{X}\mathcal{X}}$. Essentially, we are assuming that zero mean noise with a vanishingly small covariance is a good proxy for the linearization error. This introduces a bias in our estimation.

2.2.1 Simulation example for the EKF

1. Consider a nonlinear multidimensional system:

$$
\underbrace{\begin{bmatrix} x_k \\ x_k^v \\ y_k \\ y_k^v \\ \Omega_k \end{bmatrix}}_{\mathcal{X}_k} = \begin{bmatrix} 1 & \frac{\sin\Omega_k T}{\Omega_k} & 0 & -\frac{(1-\cos(\Omega_k T))}{\Omega_k} & 0 \\ 0 & \cos(\Omega_k T) & 0 & -\sin(\Omega_k T) & 0 \\ 0 & -\frac{(1-\cos(\Omega_k T))}{\Omega_k} & 1 & -\frac{\sin(\Omega_k T)}{\Omega_k} & 0 \\ 0 & \sin(\Omega_k T) & 0 & \cos(\Omega_k T) & 0 \\ 0 & 0 & 0 & 0 & 1 \end{bmatrix} \begin{bmatrix} x_{k-1} \\ x_{k-1}^v \\ y_{k-1} \\ y_{k-1}^v \\ \Omega_{k-1} \end{bmatrix} + \eta_{k-1}.
$$

(2.29)

This is, in fact, a discrete approximation of the equation of motion of an object in (x, y) plane, when the object is maneuvering with a constant but unknown turn rate Ω. The measurement is given by the range and bearing angle of the object:

$$
\mathcal{Y}_k = \begin{bmatrix} \sqrt{x_k^2 + y_k^2} \\ \text{atan2}(y_k, x_k) \end{bmatrix} + v_k,
$$

(2.30)

where atan2 is a four quadrant arctan function, implemented in MAT-LAB with a command of the same name.

2. The parameters of the system are

$$
\mathcal{X}_0 = \begin{bmatrix} 1000 & 30 & 1000 & 0 & -3 \end{bmatrix}^T,
$$

$$
\Sigma_0 = \begin{bmatrix} 200 & 0 & 0 & 0 & 0 \\ 0 & 20 & 0 & 0 & 0 \\ 0 & 0 & 200 & 0 & 0 \\ 0 & 0 & 0 & 20 & 0 \\ 0 & 0 & 0 & 0 & 0.1 \end{bmatrix},
$$

$$
Q = q \begin{bmatrix} T^3/3 & T^2/2 & 0 & 0 & 0 \\ T^2/2 & T & 0 & 0 & 0 \\ 0 & 0 & T^3/3 & T^2/2 & 0 \\ 0 & 0 & T^2/2 & T & 0 \\ 0 & 0 & 0 & 0 & 0.018T \end{bmatrix}.
$$

All the other parameters have the same values as in the previous subsection.

3. The code provided performs tasks which are the same as the previous section. These include:

(a) a state space simulation for the above system, which simulates the true measurements and the true state for 200 time-steps;

(b) implementation of the extended Kalman filter for the system to estimate the unobservable states from measurements;

(c) finding the root mean squared error *at each time step* for x_k^v and y_k^v over 100 runs and plotting these errors against time.

```
1 % EKF for a simplified constant turn model
2 clear;
3 N=200; % number of simulations
4 M=100; %number of time steps in each simulation.
5 %-- define parameter values --
6 T=0.5;  q=0.1;
7 Q= q*[T^3/3 T^2/2 0 0 0;T^2/2 T 0 0 0;
8       0 0 T^3/3 T^2/2 0; 0 0 T^2/2 T 0;
9       0 0 0 0 0.018*T];
10 Qs = sqrtm(Q);
11 Rs=[120 0;0 abs(sqrt(70))]; R=Rs*Rs';
12 x0=[1000; 30; 1000;0;-3]; P0=diag([200;20;200;20;0.1]);
13 x=[x0 zeros(5,N-1)];
14
15
16 %-- simulate the true system and run the EKF M times
17 telapsed=0;
18 for j=1:M
19 y=zeros(2,N);
20 se1=zeros(M,N); se2=se1;
21
22
23 xhat=[x0+sqrtm(P0)*randn(size(x0)) zeros(5,N-1)];
24 xhat_minus=xhat;
25 y_clean = [sqrt((x0(1))^2+(x0(3))^2);
26            atan2(x0(3),x0(1))];
27 y(:,1)= y_clean+Rs*randn(2,1);
28 Omega=[x0(5) zeros(1,N-1)];
29 % -- simulate the true system --
30 P_new=P0;
31 P_old=P0;
32 for i=1:N-1
33     sin_over_omega=sin(Omega(i)*T)/Omega(i);
34     cos_over_omega=(1-cos(Omega(i)*T))/Omega(i);
35     A=[1 sin_over_omega 0 -cos_over_omega 0;
36        0 cos(Omega(i)*T) 0 -sin(Omega(i)*T) 0;
37        0 cos_over_omega 1 sin_over_omega 0;
38        0 sin(Omega(i)*T) 0 cos(Omega(i)*T) 0;
39        0 0               0 0               1];
40
```

```
41      x(:,i+1)=A*x(:,i)+Qs*randn(5,1);
42      y_clean = [sqrt( (x(1,i+1))^2 + (x(3,i+1))^2);
43                     atan2(x(3,i+1),x(1,i+1))];
44      y(:,i+1)=y_clean+Rs*randn(2,1);
45        Omega(i+1) = x(5,i+1);
46   end
47
48   % EKF
49   tstart=tic;
50   for i=1:N-1
51      sin_over_omega=sin(Omega(i)*T)/Omega(i);
52      cos_over_omega=(1-cos(Omega(i)*T))/Omega(i);
53      A=[1   sin_over_omega  0 -cos_over_omega  0;
54         0   cos(Omega(i)*T) 0 -sin(Omega(i)*T) 0;
55         0   cos_over_omega  1 sin_over_omega   0;
56         0   sin(Omega(i)*T) 0 cos(Omega(i)*T)  0;
57         0 0                  0 0               1];
58   xhat_minus(:,i+1)= A*xhat(:,i);
59   P_old=A*P_new*A'+Q;
60   x1=xhat_minus(1,i+1); x3=xhat_minus(3,i+1);
61   C_denom1=sqrt(x1^2+x3^2);
62   C_denom2= 1+(x3^2/x1^2);
63   C = [x1/C_denom1  0 x3/C_denom1   0 0;
64    (-x3/((x1)^2))/C_denom2  0  1/(x1*C_denom2) 0 0];
65    K=P_old*C'*(C*P_old*C'+R)^(-1);
66    innov=y(:,i+1)-[sqrt(x1^2+x3^2); atan2(x3,x1)];
67    xhat(:,i+1)=xhat_minus(:,i+1)+K*innov;
68    P_C_K = P_old*C'*K';
69    P_new=P_old + K*C*P_C_K+K*R*K'-P_C_K-(P_C_K)';
70   end
71   se1(j,1:N)=(xhat(1,:)-x(1,:)).^2;
72   se2(j,1:N)=(xhat(3,:)-x(3,:)).^2;
73   telapsed=telapsed+toc(tstart);
74   end
75
76   rmse1=zeros(N,1);rmse2=rmse1;
77
78 for i=1:N
79 %-- RMSE at each timestep, averaged over M simulations
80 rmse1(i) = sqrt(sum(se1(1:M,i))/M);
81 rmse2(i) = sqrt(sum(se2(1:M,i))/M);
82 end
83 average_time = telapsed/M
84 %%-- average time per EKF run
85 close;
86
87 figure;plot(1:N, rmse1, 1:N,rmse2, '--');
88 legend( 'RMSE for x^v_k', 'RMSE for y^v_k')
```

Listing 2.2: EKF for simplified constant turn model.

A sample MATLAB code for EKF for the above target tracking problem is provided in Listing 2.2. The resulting plots of average errors are shown in Figure 2.2.

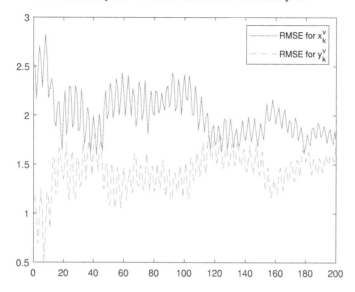

FIGURE 2.2: Root mean squared errors at each time step.

The EKF can diverge and care needs to be exercised while using it. Nevertheless, it is very commonly employed in many practical applications. Some of its common variants are described in the next section.

2.3 Important variants of the EKF

2.3.1 The iterated EKF (IEKF)

This is perhaps the most intuitive extension of the EKF. In the case of significant nonlinearities in the measurement equation, it is possible to update the state estimate and the filter gain matrix via an iterative procedure at the same time-step. The iterations are stopped when the change in the state estimate is sufficiently small in some norm, or when a maximum number of iterations is exceeded. Similar to the EKF itself, the algorithm for the IEKF consists of the time update equations $(\hat{\mathcal{X}}_{k-1|k-1}, \Sigma_{k-1|k-1}^{\mathcal{X}\mathcal{X}}) \mapsto (\hat{\mathcal{X}}_{k|k-1}, \Sigma_{k|k-1})$ and the measurement update equations $(\hat{\mathcal{X}}_{k|k-1}, \Sigma_{k|k-1}^{\mathcal{X}\mathcal{X}}) \mapsto (\hat{\mathcal{X}}_{k|k}, \Sigma_{k|k}^{\mathcal{X}\mathcal{X}})$ and this algorithm is summarized below.

As mentioned earlier, the convergence criterion could be $\|\mathcal{X}_k^{i+1} - \mathcal{X}_k^i\|_2 \le \epsilon$ for a prescribed tolerance ϵ, or it could be the maximum number of iterations. IEKF may improve the performance of the filter, at the cost of extra computation. It is *not* guaranteed to improve the performance, however; an IEKF may diverge as well.

Algorithm 5 Iterated Extended Kalman filter

$$[\hat{\mathcal{X}}_{k|k}, \Sigma_{k|k}] = \texttt{IEKF}[\mathcal{Y}_k]$$

- Initialize the filter with $\hat{\mathcal{X}}_{0|0}$ and $\Sigma_{0|0}$.

- For $k = 1 : N$

  Set $i = 1$.
  $$\hat{\mathcal{X}}_k^i = \phi_k(\hat{\mathcal{X}}_{k-1|k-1}),$$
  $$\Sigma_{k|k-1}^{\mathcal{XX}} = \nabla_{\phi_{k-1}} \Sigma_{k-1|k-1}^{\mathcal{XX}} \nabla_{\phi_{k-1}}^T + Q_k,$$

  While convergence criterion not met

  $$[\nabla_{\gamma_k}^i]_{jl} = \left[\frac{\partial \gamma_{j,k}}{\partial x_l} \right]_{x = \hat{\mathcal{X}}_k^i}$$
  $$K_k^i = \Sigma_{k|k-1}^{\mathcal{XX}} (\nabla_{\gamma_k}^i)^T (\nabla_{\gamma_k}^i \Sigma_{k|k-1}^{\mathcal{XX}} (\nabla_{\gamma_k}^i)^T + R_k)^{-1},$$
  $$\hat{\mathcal{X}}_k^{i+1} = \hat{\mathcal{X}}_k^i + K_k^i(\mathcal{Y}_k - \gamma_k(\hat{\mathcal{X}}_k^i)),$$
  $$i = i + 1,$$

  end

  $$\hat{\mathcal{X}}_{k|k} = \hat{\mathcal{X}}_k^{i+1},$$
  $$K_k = K_k^i,$$
  $$\Sigma_{k|k}^{\mathcal{XX}} = \Sigma_{k|k-1}^{\mathcal{XX}} + K_k \left(\nabla_{\gamma_k}^i \Sigma_{k|k-1}^{\mathcal{XX}} (\nabla_{\gamma_k}^i)^T + R_k \right) K_k^T$$
  $$- \Sigma_{k|k-1}^{\mathcal{XX}} (\nabla_{\gamma_k}^i)^T K_k^T - K_k \nabla_{\gamma_k}^i \Sigma_{k|k-1}^{\mathcal{XX}}.$$

- End For

2.3.2 The second order EKF (SEKF)

Another direction for possibly improving the accuracy of the EKF is to add a higher order correction term. If the functions γ, ϕ are twice continuously differentiable, one can write

$$\phi(\mathcal{X}_k, k) \approx \phi(\hat{\mathcal{X}}_{k|k}, k) + \nabla_{\phi_k}(\mathcal{X}_k - \hat{\mathcal{X}}_{k|k})$$
$$+ \frac{1}{2} \sum_{i=1}^{n_x} e_i (\mathcal{X}_k - \hat{\mathcal{X}}_{k|k})^T H_{\phi_{i,k}} (\mathcal{X}_k - \hat{\mathcal{X}}_{k|k}) + \eta_k, \qquad (2.31)$$

$$\gamma(\mathcal{X}_k, k) \approx \gamma(\hat{\mathcal{X}}_{k|k}, k) + \nabla_{\gamma_k}(\mathcal{X}_k - \hat{\mathcal{X}}_{k|k})$$
$$+ \frac{1}{2} \sum_{i=1}^{n_x} e_i (\mathcal{X}_k - \hat{\mathcal{X}}_{k|k})^T H_{\gamma_{i,k}} (\mathcal{X}_k - \hat{\mathcal{X}}_{k|k}) + v_k, \qquad (2.32)$$

where e_i denotes the ith unit vector, n_x is the dimension of the state vector and ∇_{ϕ_k}, ∇_{γ_k} are as defined earlier in Section 2.2. The Hessian matrices $H_{\gamma_{i,k}}$, $H_{\phi_{i,k}}$ are defined by

$$[H_{\phi_{i,k}}]_{jl} = \left[\frac{\partial^2 \phi_{i,k}}{\partial \mathcal{X}_j \mathcal{X}_l} \right]_{\mathcal{X} = \hat{\mathcal{X}}(k|k)},$$

$$[H_{i,k}^{\gamma}]_{jl} = \left[\frac{\partial^2 \gamma_{i,k}}{\partial \mathcal{X}_j \mathcal{X}_l} \right]_{\mathcal{X} = \hat{\mathcal{X}}(k|k)}.$$

With these approximations of the nonlinear functions, the covariance update equations are modified to account for higher order terms in the Taylor series expansion. Specifically, the new set of equations is as listed in Algorithm 6. Here, **tr** denotes the trace operator, and the formulae for covariance updates assume zero third order central moments. The SEKF requires a significant amount of extra computation as compared to the EKF, due to the computation of $2n_x$ Hessian matrices at each time-step. However, some practical applications have been reported in the literature where it performs well, see e.g., [35].

2.3.3 Divided difference Kalman filter (DDKF)

EKF requires computation of Jacobians, which might be both numerically expensive and prone to numerical errors. One way to avoid this, while still staying in the Taylor approximation framework, is to use multivariate central difference approximations for the leading order derivatives. A common way to define these approximations is in terms of operator notation (see, e.g., [117]). For a function $f(\cdot)$, we define

$$\delta_p f = f\left(\mathcal{X} + \frac{h}{2} e_p\right) - f\left(\mathcal{X} - \frac{h}{2} e_p\right), \text{ and}$$

$$\mu_p f = \frac{1}{2}\left(f\left(\mathcal{X} + \frac{h}{2} e_p\right) + f\left(\mathcal{X} - \frac{h}{2} e_p\right)\right),$$

Algorithm 6 Second order Kalman filter

$$[\hat{\mathcal{X}}_{k|k}, \Sigma_{k|k}] = \texttt{IEKF}[\mathcal{Y}_k]$$

- Initialize the filter with $\hat{\mathcal{X}}_{0|0}$ and $\Sigma_{0|0}$.

- For $k = 1 : N$

$$\hat{\mathcal{X}}_{k|k-1} = \phi_k(\hat{\mathcal{X}}_{k-1|k-1}), \tag{2.33}$$

$$\Sigma_{k|k-1}^{\mathcal{X}\mathcal{X}} = \nabla_{\phi_{k-1}} \Sigma_{k-1|k-1}^{\mathcal{X}\mathcal{X}} \nabla_{\phi_{k-1}}^T Q_k$$

$$+ \frac{1}{2} \sum_{i,j=1}^{n_x} e_i e_j^T \, \mathbf{tr} \left(H_{\phi_{i,k}} \Sigma_{k-1|k-1}^{\mathcal{X}\mathcal{X}} H_{\phi_{j,k}} \Sigma_{k-1|k-1}^{\mathcal{X}\mathcal{X}} \right), \tag{2.34}$$

$$K_k = \Sigma_{k|k-1}^{\mathcal{X}\mathcal{X}} \nabla_{\gamma_{k-1}}^T (\nabla_{\gamma_{k-1}} \Sigma_{k|k-1}^{\mathcal{X}\mathcal{X}} \nabla_{\gamma_{k-1}}^T + R_k)^{-1}, \tag{2.35}$$

$$\hat{\mathcal{X}}_{k|k} = \hat{\mathcal{X}}_{k|k-1} + K_k(\mathcal{Y}_k - \gamma(\hat{\mathcal{X}}_{k|k-1}, k)), \tag{2.36}$$

$$\Sigma_{k|k}^{\mathcal{X}\mathcal{X}} = \Sigma_{k|k-1}^{\mathcal{X}\mathcal{X}} - \Sigma_{k|k-1}^{\mathcal{X}\mathcal{X}} \nabla_{\gamma_{k-1}}^T K_k^T - K_k \nabla_{\gamma_{k-1}} \Sigma_{k|k-1}^{\mathcal{X}\mathcal{X}}$$

$$+ K_k(\nabla_{\gamma_{k-1}} \Sigma_{k|k-1}^{\mathcal{X}\mathcal{X}} \nabla_{\gamma_{k-1}}^T \tag{2.37}$$

$$+ \sum_{i,j=1}^{n_x} e_i e_j^T \mathbf{tr}(H_{\phi_{i,k}} \Sigma_{k|k-1}^{\mathcal{X}\mathcal{X}} H_{\phi_{j,k}} \Sigma_{k|k-1}^{\mathcal{X}\mathcal{X}}) R_k) K_k^T. \tag{2.38}$$

- End For

---

where e_p represents the pth unit vector and h is a time-step or scaling constant. With this notation, the first order approximation of ϕ_k based on divided differences is given by

$$\phi_k(\mathcal{X}_k) \approx \phi_k(\hat{\mathcal{X}}_{k-1|k-1}) + \frac{1}{h} \sum_{p=1}^{n} \Delta \mathcal{X}_{k,p} \left(\mu_p \delta_p \phi_k(\hat{\mathcal{X}}_{k-1|k-1}) \right),$$

where $\Delta \mathcal{X}_{k,p}$ is the perturbation in the pth element of the vector $\mathcal{X}_k$. We assume that $\mathbb{E}(\phi_k(\mathcal{X}_k)) = \phi_k(\hat{\mathcal{X}}_{k-1|k-1})$, so that $\mathbb{E}(\Delta \mathcal{X}_{k,p}) = 0$ holds for the perturbations. We then set $\hat{\mathcal{X}}_{k|k-1} = \phi_k(\hat{\mathcal{X}}_{k-1|k-1})$ as in the EKF and proceed to derive the expression for the covariance term $\Sigma_{k|k-1}^{\mathcal{X}\mathcal{X}} = \mathbb{E}(\mathcal{X}_{k|k-1} - \hat{\mathcal{X}}_{k|k-1})(\mathcal{X}_{k|k-1} - \hat{\mathcal{X}}_{k|k-1})^T$, following [117].

To achieve this, we use a transformed variable $\mathcal{X}_k^S = (S_{k-1})^{-1} \mathcal{X}_{k|k-1}$ and a function $\tilde{\phi}_k(\mathcal{X}_k^S) := \phi_k(S_{k-1} \mathcal{X}_k) = \phi_k(\mathcal{X}_{k|k-1})$. Here S_{k-1} is a square Cholesky factor of $\Sigma_{k-1|k-1}^{\mathcal{X}\mathcal{X}}$. Now, the first order approximation for $\tilde{\phi}_k$ is

$$\tilde{\phi}_k(\mathcal{X}_k^S) \approx \hat{\mathcal{X}}_{k|k-1} + \frac{1}{h} \sum_{p=1}^{n} \Delta \mathcal{X}_{k,p}^S \left(\mu_p \delta_p \tilde{\phi}_k(\hat{\mathcal{X}}^S{}_{k-1|k-1}) \right), \qquad (2.39)$$

where $\hat{\mathcal{X}}_{k-1|k-1}^S = (S_{k-1})^{-1} \hat{\mathcal{X}}_{k-1|k-1}$ and $\Delta \mathcal{X}^S$ is a zero mean perturbation in the transformed variable $\mathcal{X}^S$. Note that

$$\mathbb{E}(\Delta \mathcal{X}_{k,p}^S)(\Delta \mathcal{X}_{k,p}^S)^T = S_{k-1}^{-1} \mathbb{E}(\Delta \mathcal{X}_{k,p})(\Delta \mathcal{X}_{k,p})^T (S_{k-1}^{-1})^T = I, \qquad (2.40)$$

which is the reason for our choice of linear transformation. Using (2.40) and (2.39) leads to

$$\mathbb{E}(\mathcal{X}_{k|k-1} - \phi_k(\hat{\mathcal{X}}_{k-1|k-1}))(\mathcal{X}_{k|k-1} - \phi_k(\hat{\mathcal{X}}_{k-1|k-1}))^T$$
$$= \frac{1}{h} \sum_{p=1}^{n} \left(\mu_p \delta_p \tilde{\phi}_k(\hat{\mathcal{X}}^S{}_{k-1|k-1}) \right) \left(\mu_p \delta_p \tilde{\phi}_k(\hat{\mathcal{X}}^S{}_{k-1|k-1}) \right)^T,$$

which after simplification gives the necessary expression for the predicted covariance matrix:

$$\Sigma_{k|k-1}^{\mathcal{X}\mathcal{X}} = \frac{1}{4h^2} \sum_{p=1}^{n} \left\{ \phi_k(\hat{\mathcal{X}}_{k|k-1} + h S_{k-1,p}) - \phi_k(\hat{\mathcal{X}}_{k|k-1} - h S_{k-1,p}) \right\}$$
$$\times \left\{ \phi_k(\hat{\mathcal{X}}_{k|k-1} + h S_{k-1,p}) - \phi_k(\hat{\mathcal{X}}_{k|k-1} - h S_{k-1,p}) \right\}^T + Q_k, \qquad (2.41)$$

where $S_{k-1,p}$ is the pth column of the Cholesky factor S_{k-1}. Following a similar logic, we have

$$\Sigma_{k|k-1}^{yy} = \frac{1}{4h^2} \sum_{p=1}^{n} \left\{ \gamma_k(\hat{\mathcal{X}}_{k|k-1} + hS_{k-1,p}) - \gamma_k(\hat{\mathcal{X}}_{k|k-1} - hS_{k-1,p}) \right\}$$

$$\times \left\{ \gamma_k(\hat{\mathcal{X}}_{k|k-1} + hS_{k-1,p}) - \gamma_k(\hat{\mathcal{X}}_{k|k-1} - hS_{k-1,p}) \right\}^T + R_k \quad \text{and} \tag{2.42}$$

$$\Sigma_{k|k-1}^{xy} = \frac{1}{2h} \sum_{p=1}^{n} S_{k-1,p} \left\{ \gamma_k(\hat{\mathcal{X}}_{k|k-1} + hS_{k-1,p}) - \gamma_k(\hat{\mathcal{X}}_{k|k-1} - hS_{k-1,p}) \right\}^T. \tag{2.43}$$

Combining the above with the covariance based definition of Kalman gain, we can then write the steps for the update $(\hat{\mathcal{X}}_{k-1|k-1}, \Sigma_{k-1|k-1}^{xx}) \mapsto (\hat{\mathcal{X}}_{k|k}, \Sigma_{k|k}^{xx})$ at time step k as follows: According to [117], h is usually chosen to be $\sqrt{3}$ since

---

Algorithm 7 Divided difference Kalman filter

$$[\hat{\mathcal{X}}_{k|k}, \Sigma_{k|k}] = \text{DDKF}[\mathcal{Y}_k]$$

- Initialize the filter with $\hat{\mathcal{X}}_{0|0}$ and $\Sigma_{0|0}$.

- For $k = 1 : N$

    1. Find a square Cholesky factor S_{k-1} for $\Sigma_{k-1|k-1}^{xx}$.

    2. Evaluate $\Sigma_{k|k-1}^{xx}, \Sigma_{k|k-1}^{yy}$ and $\Sigma_{k|k-1}^{xy}$ according to (2.41), (2.42) and (2.43), respectively.

    3. Set Kalman gain $K_k = \Sigma_{k|k-1}^{xy} (\Sigma_{k|k-1}^{yy})^{-1}$.

    4. Evaluate $\hat{\mathcal{X}}_{k|k}$ and $\Sigma_{k|k}^{xx}$ by

    $$\hat{\mathcal{X}}_{k|k} = \phi_k(\hat{\mathcal{X}}_{k-1|k-1}) + K_k(\mathcal{Y}_k - \gamma_k(\hat{\mathcal{X}}_{k|k-1})),$$
    $$\Sigma_{k|k}^{xx} = \Sigma_{k|k-1}^{xx} - K_k \Sigma_{k|k-1}^{yy} K_k^T. \tag{2.44}$$

- End For

---

it improves matching of fourth order moments (at least in the scalar case). This filter avoids computation of Jacobians entirely, replacing them with extra function evaluations at a set of deterministic sample points $\hat{\mathcal{X}}_{k|k-1} \pm hS_{k-1,p}$, $p = 1, 2, \cdots, n$. In a sense, this method may be seen as a bridge between the EKF and the methods in subsequent chapters which rely on extra function evaluations at a set of appropriately - but deterministically - chosen sample points.

The covariance computation described above can also be extended for a second order central difference approximation of functions; the second order version of the divided difference filter is described in [117] and [91] for example. [158] discusses an adaptive version of the second order filter. The same references also discuss a square root implementation of the above algorithm via Householder transformation.

2.3.4 MATLAB-based filtering exercises

1. Repeat the numerical experiment in Section 2.2.1 with different values of parameters; e.g., see how changing the size of noise affects the errors.

2. Implement IEKF, SEKF and DDKF for the same simulation example, and compare the performance of the three filters in terms of rmse at each time step.

2.4 Alternative approaches towards nonlinear filtering

EKF and its variants approach nonlinear filtering by linearizing the system dynamics and then using the Kalman filter. This approach has severe limitations; the system dynamics may be difficult or impossible to linearize (e.g., due to non-differentiability) and it may not be possible to compute Jacobians in closed-form. There are two main approaches to nonlinear filtering which avoid linearization either entirely or at least at the covariance computation stage. Both the approaches are based on the idea of propagating an approximate distribution of the state via a set of samples.

- In the *particle* filtering (PF) or sequential Monte Carlo filtering approach, the posterior distribution of state is propagated by sampling from a proposal density (which may or may not be the transition density itself) and then adjusting the corresponding probability weights to be consistent with the observations, using Bayes' rule. This technique was outlined earlier in Chapter 1.

- In Unscented Kalman filtering (UKF) or the sigma point filtering approach, a set of deterministic chosen samples and associated weights represent the distribution of the state. These are propagated through the state dynamics and are used to calculate the mean and the covariance matrix of the updated state using a Kalman filter like formula. In essence, the terms in Eq. (2.7) are computed numerically using a set of samples and weights, instead of being computed using the Gaussian conditional probability formula. Unlike the PF, the samples are chosen via a deterministic rule in the UKF, e.g., to match a given covariance matrix. Further, the weights may not be valid probability weights. One may also view the UKF as a method of recursively approximating the integrations involved in computing the means and the covariance matrices in (2.7). Cubature Kalman filters (CKF) and cubature quadrature Kalman filters (CQKF) are somewhat more sophisticated approaches which use error controlled integration for computing the same quantities. Similar to the UKF, CKF and CQKF do not involve random sampling. We will discuss the UKF, the CKF and the CQKF in detail in subsequent chapters.

2.5   Summary

This chapter looked at the derivation and the properties of the linear Kalman filter, and the approaches towards nonlinear filtering based on local linearization of the system dynamics at each time-step. While these approaches have often been successful in practice, the last couple of decades have seen major advances in approaches which do not depend on explicit system linearization. We will look at some of these filtering approaches in detail in subsequent chapters.

Chapter 3

Unscented Kalman filter

3.1   Introduction 51
3.2   Sigma point generation 52
3.3   Basic UKF algorithm 54
    3.3.1   Simulation example for the unscented Kalman Filter ... 56
3.4   Important variants of the UKF 60
    3.4.1   Spherical simplex unscented transformation 60
    3.4.2   Sigma point filter with $4n + 1$ points 61
    3.4.3   MATLAB-based filtering exercises 64
3.5   Summary 64

3.1   Introduction

We start by recalling that, if $\mathcal{X}_k \mid \mathcal{Y}_{k-1}, \mathcal{Y}_k \mid \mathcal{Y}_{k-1}$ are jointly Gaussian, the conditional mean $\hat{\mathcal{X}}_{k|k} = \mathbb{E}(\mathcal{X}_k \mid \mathcal{Y}_k)$ and the conditional variance $\Sigma_{k|k}^{\mathcal{X}\mathcal{X}} = \mathrm{Var}(\mathcal{X}_k \mid \mathcal{Y}_k)$ are given by

$$\mathbb{E}(\mathcal{X}_k \mid \mathcal{Y}_k) = \mathbb{E}(\mathcal{X}_k \mid \mathcal{Y}_{k-1}) + \Sigma_{k|k-1}^{\mathcal{X}\mathcal{Y}}(\Sigma_{k|k-1}^{\mathcal{Y}\mathcal{Y}})^{-1}(\mathcal{Y}_k - \mathbb{E}(\mathcal{Y}_k \mid \mathcal{Y}_{k-1})), \tag{3.1}$$

$$\text{and } \mathrm{Var}(\mathcal{X}_k \mid \mathcal{Y}_k) = \Sigma_{k|k-1}^{\mathcal{X}\mathcal{X}} - \Sigma_{k|k-1}^{\mathcal{X}\mathcal{Y}}(\Sigma_{k|k-1}^{\mathcal{Y}\mathcal{Y}})^{-1}(\Sigma_{k|k-1}^{\mathcal{X}\mathcal{Y}})^{T}, \tag{3.2}$$

where $\Sigma_{k|k-1}^{\mathcal{X}\mathcal{Y}}$, $\Sigma_{k|k-1}^{\mathcal{Y}\mathcal{Y}}$ and $\Sigma_{k|k-1}^{\mathcal{X}\mathcal{X}}$ are the cross-covariance matrix between $\mathcal{X}_k \mid \mathcal{Y}_{k-1}$ and $\mathcal{Y}_k \mid \mathcal{Y}_{k-1}$, the auto-covariance matrix of $\mathcal{Y}_k \mid \mathcal{Y}_{k-1}$ and the auto-covariance matrix of $\mathcal{X}_k \mid \mathcal{Y}_{k-1}$, respectively. In the EKF and related methods described in the last chapter, the matrices are computed after linearizing the dynamics. An alternative approach, followed in the unscented Kalman filter (UKF) and related methods, is to construct an approximate discrete 'distribution' of $\mathcal{X}_k \mid \mathcal{Y}_{k-1}$ and $\mathcal{Y}_k \mid \mathcal{Y}_{k-1}$ with a small number of points and then propagate it using the equations above. The distribution consists of a set of deterministic sample points (or *sigma points*) and associated weights, although the weights may not all be positive and hence it may not be a valid probability distribution. As we still use (3.1)-(3.2), we are still operating in a Gaussian framework; however, we can avoid evaluating the Jacobians at each time-step. If the EKF is seen as a 'linearize first, then evaluate covariance matrices' approach, the UKF may be seen as an 'evaluate covariance matrices first, then linearize' approach.

The rest of the chapter is organized as follows. Before we describe some of the standard UKF algorithms, we first take a close look at the properties of sigma point generation algorithms based on computing a square root of the covariance matrix in the next section. A basic UKF algorithm is described in Section 3.3, and its significant extensions are described in Section 3.4. Section 3.5 summarises the chapter.

3.2 Sigma point generation

Most sigma point generation algorithms are based on decomposition of the associated covariance matrix. Suppose that, at time-step k, we have the conditional mean $\hat{\mathcal{X}}_{k|k}$ and the conditional covariance $\Sigma_{k|k}^{\mathcal{X}\mathcal{X}}$ available. With n being the state dimension, we can generate a set of $2n + 1$ discrete vectors $\mathcal{X}_{k|k}^i$ and associated weights w^i to match these two moments as follows:

- Find a real square root matrix S_k such that $S_k S_k^T = \Sigma_{k|k}^{\mathcal{X}\mathcal{X}}$. Real square roots for symmetric positive definite matrices are not unique, and we choose a square root matrix which is easy to compute, e.g., the Cholesky factor of $\Sigma_{k|k}^{\mathcal{X}\mathcal{X}}$ (see [70] for details on Cholesky factorisation). Let S_k^i be the ith column of S_k.

- Choose the sigma points $\mathcal{X}_{k|k}^i$ as

$$
\begin{aligned}
\mathcal{X}_{k|k}^0 &= \hat{\mathcal{X}}_{k|k}, \\
\mathcal{X}_{k|k}^i &= \hat{\mathcal{X}}_{k|k} + \sqrt{n+\lambda}S_k^i,\ i = 1, 2, \cdots, n, \\
\mathcal{X}_{k|k}^i &= \hat{\mathcal{X}}_{k|k} - \sqrt{n+\lambda}S_k^i,\ i = n+1, n+2, \cdots, 2n, \quad (3.3)
\end{aligned}
$$

where λ is a scaling parameter.

- Choose the weights w_k^i as

$$
\begin{aligned}
w^0 &= \frac{\lambda}{n+\lambda}, \\
w^i &= \frac{1}{2(n+\lambda)},\ i = 1, 2, \cdots, 2n. \quad (3.4)
\end{aligned}
$$

With the above definitions, one can easily compute the mean and the covariance matrix of the discrete 'distribution' defined by $\mathcal{X}_{k|k}^i$ as support points and w^i as 'probability' weights as

$$\sum_{i=0}^{2n} w_{k|k}^i \mathcal{X}_{k|k}^i = \hat{\mathcal{X}}_{k|k}, \text{and}$$

$$\sum_{i=0}^{2n} w_{k|k}^i (\mathcal{X}_{k|k}^i - \hat{\mathcal{X}}_{k|k})(\mathcal{X}_{k|k}^i - \hat{\mathcal{X}}_{k|k})^T = \frac{1}{2(n+\lambda)} \sum_{i=1}^{2n} (n+\lambda) S_k^i (S_k^i)^T = \Sigma_{k|k}^{\mathcal{X}\mathcal{X}}.$$

$$(3.5)$$

Alternatively, if $q(x)$ is the continuous multivariate probability density for $\mathcal{X}_{k|k}$, the above equations may be seen as approximations of multivariate integrals:

$$\sum_{i=0}^{2n} w_{k|k}^i \mathcal{X}_{k|k}^i = \int \mathcal{X}_{k|k} q(x) dx \quad and$$

$$\sum_{i=0}^{2n} w_{k|k}^i (\mathcal{X}_{k|k}^i - \hat{\mathcal{X}}_{k|k})(\mathcal{X}_{k|k}^i - \hat{\mathcal{X}}_{k|k})^T = \int (\mathcal{X}_{k|k} - \hat{\mathcal{X}}_{k|k})(\mathcal{X}_{k|k} - \hat{\mathcal{X}}_{k|k})^T q(x) dx.$$

This latter interpretation is relevant in subsequent chapters when we look at cubature Kalman filters. Clearly, $\sum_{i=0}^{2n} w_k^i = 1$. If, in addition, $\lambda > 0$, $\{w^i\}$ define valid probability weights. In the literature, the weights are often allowed to be negative, with $\lambda < 0$. In particular, [85] and subsequent publications recommend using $\lambda = 3 - n$. This makes the interpretation of sigma points (either from a probabilistic point of view or from a numerical quadrature point of view) rather difficult. We will focus on the case of positive λ in the ensuing discussion in this section. In subsequent sections, we will also look at the use of such free parameters to match other useful properties of the underlying distributions.

To see what the above moment matching achieves, consider a real function $f_1 : \mathbb{R}^n \mapsto \mathbb{R}$ as a function of the following form:

$$f(x) = \sum_{i,j=1}^{n} a_{ij} x_i x_j + \sum_{i=1}^{n} (b_i x_i),$$

where a_{ij}, b_i, are arbitrary real constants. Let $\mathcal{X}_{k|k}$ be a vector-valued random variable on continuous support with mean $\hat{\mathcal{X}}_{k|k}$ and covariance matrix $\Sigma_{k|k}^{\mathcal{X}\mathcal{X}}$, and let X be a vector-valued random variable defined by discrete support points $\mathcal{X}_{k|k}^i$ and probability weights $w_{k|k}^i$, which match the first two moments exactly as described above. If $\mathbb{E}_\mathcal{X}$ and $\mathbb{E}_X$ represent the mean according to the distribution of $\mathcal{X}_k$ and X, respectively, it can be easily seen that $\mathbb{E}_\mathcal{X}(f(\mathcal{X})) = \mathbb{E}_X(f(X))$. If, in addition, $a_{ij} = 0$, $\mathbb{E}_\mathcal{X}(f(\mathcal{X})^2) = \mathbb{E}_X(f(X)^2)$. Thus, for systems which have locally at most quadratic dynamics, the distribution defined by sigma points and the weights above gives the correct $\mathbb{E}(f(\mathcal{X}_{k|k}))$. For systems which are locally 'almost linear' with very small a_{ij}

values, the covariance matrix can also be matched, at least approximately. This is achieved without random sampling, without computation of any Jacobians and with only a small number of support points. This computational simplicity is the main advantage of the UKF and related algorithms. It is incorrectly stated in many papers that UKF calculates the covariance matrix correctly up to second order; i.e., the covariance matching should work for $a_{ij} \neq 0$ (see, e.g., [87]). However, [63] has illustrated that this is not true and has provided a counterexample.

The next section outlines the original UKF algorithm which uses the above sigma point generation method.

3.3   Basic UKF algorithm

We consider a nonlinear system with zero mean, uncorrelated additive noise:

$$\mathcal{X}_{k+1} = \phi_k(\mathcal{X}_k) + \eta_k,$$
$$\mathcal{Y}_k = \gamma_k(\mathcal{X}_k) + v_k,$$

where the functions ϕ_k, γ_k and noise variables η_k, v_k have the same properties as in Section 1.7.2. The algorithm for unscented Kalman filtering for this system is described below.

- Initialize $\hat{\mathcal{X}}_{0|0} = \mathbb{E}(\mathcal{X}_0), \Sigma_{0|0}^{\mathcal{X}\mathcal{X}} = \mathbb{E}(\mathcal{X}_0 - \hat{\mathcal{X}}_{0|0})(\mathcal{X}_0 - \hat{\mathcal{X}}_{0|0})^T$. $S_{0|0}S_{0|0}^T = \Sigma_{0|0}^{\mathcal{X}\mathcal{X}}$.

- At each time $k > 0$:

  1. For $i = 1, 2, \cdots, n$, generate

  $$\mathcal{X}_{k|k}^i = \hat{\mathcal{X}}_{k|k} + \sqrt{N + \lambda}S_k^i,$$
  $$\mathcal{X}_{k|k}^{(n+i)} = \hat{\mathcal{X}}_{k|k} - \sqrt{n + \lambda}S_k^i,$$
  $$\mathcal{X}_{k|k}^0 = \hat{\mathcal{X}}_{k|k}.$$

  2. Generate the corresponding probability weights $w_{k|k}^i$ as in equation (3.4).

3. Carry out **time update** :

$$\mathcal{X}^i_{k+1|k} = \phi_k(\mathcal{X}^i_{k|k}),$$

$$\hat{\mathcal{X}}_{k+1|k} = \sum_{i=0}^{2n} w^i \mathcal{X}^i_{k+1|k},$$

$$\Sigma^{\mathcal{X}\mathcal{X}}_{k+1|k} = \sum_{i=0}^{2n} w^i (\mathcal{X}^i_{k+1|k} - \hat{\mathcal{X}}_{k+1|k})(\mathcal{X}^i_{k+1|k} - \hat{\mathcal{X}}_{k+1|k})^T + Q,$$

$$\mathcal{Y}^i_{k+1|k} = \gamma_k(\mathcal{X}^i_{k+1|k}),$$

$$\hat{\mathcal{Y}}_{k+1|k} = \sum_{i=0}^{2n} w^i \mathcal{Y}^i_{k+1|k}.$$

4. Once the measurement $\mathcal{Y}_{k+1}$ is received, carry out the **measurement update** :

$$\Sigma^{\mathcal{Y}\mathcal{Y}}_{k+1|k} = \sum_{i=0}^{2n} w^i (\mathcal{Y}^i_{k+1|k} - \hat{\mathcal{Y}}_{k+1|k})(\mathcal{Y}^i_{k+1|k} - \hat{\mathcal{Y}}_{k+1|k})^T + R,$$

$$\Sigma^{\mathcal{X}\mathcal{Y}}_{k+1|k} = \sum_{i=0}^{2n} w^i (\mathcal{X}^i_{k+1|k} - \hat{\mathcal{X}}_{k+1|k})(\mathcal{Y}^i_{k+1|k} - \hat{\mathcal{Y}}_{k+1|k})^T,$$

$$K_{k+1} = \Sigma^{\mathcal{X}\mathcal{Y}}_{k+1|k} (\Sigma^{\mathcal{Y}\mathcal{Y}}_{k+1|k})^{-1},$$

$$\hat{\mathcal{X}}_{k+1|k+1} = \hat{\mathcal{X}}_{k+1|k} + K_{k+1}(\mathcal{Y}_{k+1} - \hat{\mathcal{Y}}_{k+1|k}),$$

$$\Sigma^{\mathcal{X}\mathcal{X}}_{k+1|k+1} = \Sigma^{\mathcal{X}\mathcal{X}}_{k+1|k} - K_{k+1}\Sigma^{\mathcal{Y}\mathcal{Y}}_{k+1|k}K^T_{k+1}.$$

- Set $k := k + 1$.

Note that, unlike the EKF, the initial covariance $\Sigma^{\mathcal{X}\mathcal{X}}_{0|0}$ should not be assumed to be too large, as the initial sigma points will be too far from the mean.

As the steps above illustrate, the UKF doesn't require evaluation of Jacobians (unlike the EKF) and it doesn't require random sampling (unlike the PF). It may thus be seen as a compromise between the EKF and the PF. As [63] demonstrates, it is closer in principle to the EKF (specifically, an implementation of the EKF with finite difference approximation of the Jacobians) since both the EKF and the UKF assume the latent and measurement variables to be jointly Gaussian and use the same key relationship (2.7) to derive the filter. This relationship is also used in the cubature Kalman filter (CKF), to be discussed later, and lies at the heart of all filtering algorithms which use deterministic sampling points.

Finally, note that the above algorithm can easily be extended to systems with non-additive noise, by augmenting the state vector with the noise vector and considering a Cholesky factor of their combined covariance matrix. This may be seen as an advantage over the EKF.

3.3.1   Simulation example for the unscented Kalman filter

1. Consider the same nonlinear multidimensional system as in Section 2.2.1:

$$
\underbrace{\begin{bmatrix} x_k \\ x_k^v \\ y_k \\ y_k^v \\ \Omega_k \end{bmatrix}}_{\mathcal{X}_k} = \begin{bmatrix} 1 & \frac{\sin \Omega_k T}{\Omega_k} & 0 & -\frac{(1-\cos(\Omega_k T))}{\Omega_k} & 0 \\ 0 & \cos(\Omega_k T) & 0 & -\sin(\Omega_k T) & 0 \\ 0 & -\frac{(1-\cos(\Omega_k T))}{\Omega_k} & 1 & -\frac{\sin(\Omega_k T)}{\Omega_k} & 0 \\ 0 & \sin(\Omega_k T) & 0 & \cos(\Omega_k T) & 0 \\ 0 & 0 & 0 & 0 & 1 \end{bmatrix} \begin{bmatrix} x_{k-1} \\ x_{k-1}^v \\ y_{k-1} \\ y_{k-1}^v \\ \Omega_{k-1} \end{bmatrix} + \eta_{k-1}.
$$

$$(3.6)$$

As before, the measurement is given by the range and bearing angle of the object:

$$
\mathcal{Y}_k = \begin{bmatrix} \sqrt{x_k^2 + y_k^2} \\ \mathrm{atan2}(y_k, x_k) \end{bmatrix} + v_k, \tag{3.7}
$$

where atan2 is a four quadrant arctan function, implemented in MAT-LAB with a command of the same name.

2. The parameters of the system are the same as before, and are repeated here for completeness:

$$
\mathcal{X}_0 = \begin{bmatrix} 1000 & 30 & 1000 & 0 & -3 \end{bmatrix}^T,
$$

with the initial covariance

$$
\Sigma_0 = \begin{bmatrix} 200 & 0 & 0 & 0 & 0 \\ 0 & 20 & 0 & 0 & 0 \\ 0 & 0 & 200 & 0 & 0 \\ 0 & 0 & 0 & 20 & 0 \\ 0 & 0 & 0 & 0 & 0.1 \end{bmatrix}.
$$

All the other parameters, including Q, R matrices have the same values as in the previous chapter.

3. The code provided performs tasks which are the same as the previous chapter, this time using the unscented Kalman filter. These tasks include:

   (a) a state space simulation for the above system, which simulates the true measurements and the true state for 200 time-steps;

   (b) implementation of the UKF (with $\lambda = 3 - n$) for the system to estimate the unobservable states from measurements;

(c) finding the root mean squared error *at each time step* for x_k^v and y_k^v over 100 runs and plotting these errors against time.

```
1 % UKF for simplified constant turn rate model
2 clear;
3 N=200; % number of simulations
4 M=100; %number of time steps in each simulation.
5 %— define parameter values —
6 T=0.5;   q=0.1;
7 Q= q*[T^3/3 T^2/2 0 0 0; T^2/2 T 0 0 0;
8        0 0 T^3/3 T^2/2 0; 0 0 T^2/2 T 0;
9        0 0 0 0 0.018*T];
10   Rs=[120 0;0 abs(sqrt(70))]; R=Rs*Rs';
11 x0=[1000; 30; 1000;0;-3]; P0=diag([200;20;200;20;0.1]);
12 Omega=[x0(5); zeros(N-1,1)];
13 x=[x0 zeros(5,N-1)];
14
15  %— simulate the true system and run the UKF M times
16 telapsed=0;
17 for j=1:M
18 y=zeros(2,N);
19 se1=zeros(M,N); se2=se1;
20
21
22 xhat=[x0+sqrtm(P0)*randn(size(x0)) zeros(5,N-1)];
23  xh_m=xhat;
24  y_clean = [sqrt((x0(1))^2+(x0(3))^2);
25             atan2(x0(3),x0(1))];
26  y(:,1)= y_clean+Rs*randn(2,1);
27  Omega=[x0(5) zeros(1,N-1)];
28 % —— simulate the true system ——
29 P_new=P0;
30 P_old=P0;
31 for i=1:N-1
32     sin_over_omega=sin(Omega(i)*T)/Omega(i);
33     cos_over_omega=(1-cos(Omega(i)*T))/Omega(i);
34     A=[1  sin_over_omega  0 -cos_over_omega  0;
35        0  cos(Omega(i)*T) 0 -sin(Omega(i)*T) 0;
36        0  cos_over_omega  1 sin_over_omega   0;
37        0  sin(Omega(i)*T) 0 cos(Omega(i)*T)  0;
38        0 0              0 0               1];
39
40     x(:,i+1)=A*x(:,i)+Qs*randn(5,1);
41     y_clean = [sqrt( (x(1,i+1))^2 + (x(3,i+1))^2);
42                atan2(x(3,i+1),x(1,i+1))];
43     y(:,i+1)=y_clean+Rs*randn(2,1);
44        Omega(i+1) = x(5,i+1);
45 end
46
47 %UKF*******
48 tstart=tic;
49  lambda=-1;
50 xhat=[x0+sqrtm(P0)*randn(size(x0)) zeros(5,N-1)];
51 xsigma = zeros(5,2*5+1); %sigma points
52 Psq=sqrtm(P0);
53 sqlambda= sqrt(5+lambda);
```

```
54 for i=1:5
55    xsigma(:,i) = xhat(:,1)+sqlambda*Psq(:,i);
56    xsigma(:,6+i)=xhat(:,1)-sqlambda*Psq(:,i);
57 end
58 %initialize variables.
59 xsigma(:,6) = xhat(:,1);
60 xs_m=xsigma;
61 ys_m=zeros(2,11);
62 yh_m=zeros(2,N);
63 xh_m=zeros(5,N);
64 w= [ (1/(2*(5+lambda)))*ones(5,1);
65      lambda/(5+lambda);
66      (1/(2*(5+lambda)))*ones(5,1)];
67
68 for k=1:11
69    ys_m(:,k)=[sqrt((xs_m(3,k))^2 + (xs_m(1,k))^2);
70               atan2(xs_m(3,k),xs_m(1,k))];
71    yh_m(:,1) = yh_m(:,1)+w(k)*ys_m(:,k);
72 end
73
74
75 for i=1:N-1
76
77 for k=1:11 %%time update
78    xs_m(:,k) = A*xsigma(:,k);
79 end
80 for k=1:11
81 xh_m(:,i+1) = xh_m(:,i+1)+w(k)*xs_m(:,k);
82 end
83 P_old=zeros(size(P0));
84 x_innov=zeros(size(xs_m));
85 for k=1:11
86 x_ki=(xs_m(:,k)-xh_m(:,i+1))*(xs_m(:,k)-xh_m(:,i+1))';
87 P_old=P_old+w(k)*x_ki;
88 ys_m(:,k)=[sqrt((xs_m(3,k))^2 + (xs_m(1,k))^2);
89               atan2(xs_m(3,k),xs_m(1,k))];
90 yh_m(:,i+1) = yh_m(:,i+1)+w(k)*ys_m(:,k);
91 end
92 P_old=P_old+Q;
93 y_innov = zeros(size(ys_m));
94 Pxy=zeros(5,2);
95 Pyy=zeros(2,2);
96 for k=1:11 %%measurement update.
97    y_innov(:,k)= ys_m(:,k)-yh_m(:,i+1);
98    Pxy=Pxy+w(k)*(xs_m(:,k)-xh_m(:,i+1))*y_innov(:,k)';
99    Pyy=Pyy+w(k)*y_innov(:,k)*y_innov(:,k)';
100 end
101 Pyy=Pyy+R;
102 K=Pxy*Pyy^(-1);
103 xhat(:,i+1)= xh_m(:,i+1)+K*(y(:,i+1)-yh_m(:,i+1));
104 P_new = P_old - K*Pyy*K';
105 Psq=chol(P_new);
106 for k=1:5
107    xsigma(:,k) = xhat(:,i+1)+sqlambda*Psq(:,k);
108    xsigma(:,6+k)=xhat(:,i+1)-sqlambda*Psq(:,k);
109 end
110 xsigma(:,6) = xhat(:,i+1);
```

```
111 end
112 se1(j,1:N)=(xhat(1,:)-x(1,:)).^2;
113 se2(j,1:N)=(xhat(3,:)-x(3,:)).^2;
114
115 telapsed=telapsed+toc(tstart);
116 end
117
118 rmse1=zeros(N,1);rmse2=rmse1;
119
120 for i=1:N
121 %-- RMSE at each timestep, averaged over M simulations
122 rmse1(i) = sqrt(sum(se1(1:M,i))/M);
123 rmse2(i) = sqrt(sum(se2(1:M,i))/M);
124 end
125 average_time = telapsed/M
126 %%--- average time per UKF run
127 close;
128
129 figure;plot(1:N, rmse1, 1:N,rmse2, '-- ');
130 legend( 'RMSE for x^v_k ', 'RMSE for y^v_k ')
```

Listing 3.1: UKF for simplified constant turn model.

A sample MATLAB code for UKF is provided in Listing 3.1. The resulting plots of average errors are shown in Figure 3.1.

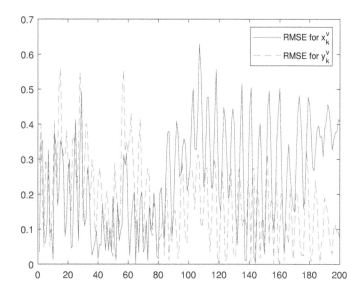

FIGURE 3.1: Root mean squared errors at each time step.

3.4 Important variants of the UKF

The algorithm outlined in this section uses $2n + 1$ sigma points. In some applications, the state dimension is high relative to computing power available, and a smaller number of points is desirable. It is also possible in some other applications that a larger number of sigma points are desirable, perhaps to match some other statistical properties of distribution. The next two subsequent subsections describe key variants of the sigma point generation step of the UKF which allow for a smaller or a larger number of sigma points.

3.4.1 Spherical simplex unscented transformation

In spherical simplex based transformation, there are $n + 1$ sigma points on a hypersphere centred at the origin and one sigma point at the origin. Thus the total number of sigma points is $n + 2$, instead of $2n + 1$ as in the algorithm earlier. The weight of the point at the origin is often taken as zero, in which case there are only $n + 1$ points. A smaller number of sigma points leads to a corresponding reduction in computational time. Suppose that $\Sigma_{k|k}^{\mathcal{X}\mathcal{X}} = S_k S_k^T$ and $\hat{\mathcal{X}}_{k|k}$ are given, as in the case of the UKF in the previous section. The algorithm to generate sigma points in the spherical simplex algorithm is given below, with the time subscript k omitted for notational simplicity:

1. Choose $0 < w^0 < 1$.

2. Choose weight sequence:
$$w^i = (1-w^0)/n+1, \ i = 1, 2, \cdots, n.$$

3. Initialize sigma point vector sequence as
$$\mathcal{X}^{01} = 0, \ \mathcal{X}^{11} = -1/\sqrt{2w^1}, \mathcal{X}^{21} = 1/\sqrt{2w^1}.$$

4. Expand the sigma point vector sequence for $i = 1, 2, \cdots, n+1, j = 1, 2, \cdots, n$ according to
$$\mathcal{X}^{ij} = \begin{bmatrix} \mathcal{X}^{0(j-1)} \\ 0 \end{bmatrix} \text{ for } i = 0$$
$$= \begin{bmatrix} \mathcal{X}^{i(j-1)} \\ -\dfrac{1}{\sqrt{j(j+1)w^1}} \end{bmatrix} \text{ for } i = 1, \cdots, j,$$
$$= \begin{bmatrix} 0^{j-1} \\ \dfrac{j}{\sqrt{j(j+1)w^1}} \end{bmatrix} \text{ for } i = j+1,$$

where 0^{j-1} is a vector in $\mathbb{R}^{j-1}$ with all zero elements.

5. Now set

$$\mathcal{X}^j = \hat{\mathcal{X}}_{k|k} + S_k \mathcal{X}_n^j.$$

More details on this algorithm may be found in [86] and [84], which also provide heuristics for choosing w^0 and numerical examples of its use for filtering in high dimensional nonlinear systems. Note that, unlike the traditional UKF, spherical simplex sigma points have positive weights and they define a valid probability distribution.

3.4.2 Sigma point filter with $4n+1$ points

As mentioned earlier, some applications might benefit from a higher number of sigma points in terms of improved accuracy. Further, one may wish to represent the properties of the underlying distribution in a more realistic fashion. The algorithm in the previous subsection leads to equal weights for n points, while the traditional UKF algorithm leads to negative weights in general. We describe here in some detail a recent algorithm proposed in [128] which allows generation of $4n + 1$ sigma points which match the first two moments and also guarantee than the mode of the multivariate distribution coincides with its mean, as is the case with the Gaussian distribution.

The following steps represent the proposed new sigma point transformation for a given covariance matrix Σ and a given mean $\hat{\mathcal{X}}$:

- Find any real S such that $SS^T = \Sigma^{\mathcal{X}\mathcal{X}}$, and S^i and $\Sigma_i^{\mathcal{X}\mathcal{X}}$ denote the ith column of S and $\Sigma^{\mathcal{X}\mathcal{X}}$, respectively.

- Calculate $\alpha_i = \frac{|<\hat{\mathcal{X}},\Sigma_i^{\mathcal{X}\mathcal{X}}>|}{\|\hat{\mathcal{X}}\|_2 \|\Sigma_i^{\mathcal{X}\mathcal{X}}\|_2}$.

- Choose a real constant $m \in (0.5, 1)$.

- Choose a real constant β such that $\beta > \{\frac{1}{4}\max(m\alpha_i) - \frac{1}{2}\sum_{i=1}^n \alpha_i\}$.

Now define a discrete distribution for the random variable $\mathbb{Z}$ as

$$\mathbb{P}(\mathbb{Z} = \mu) = 1 - \frac{\sum_{i=1}^n \alpha_i}{2(\sum_{i=1}^n \alpha_i + \beta)},$$

$$\mathbb{P}\left(\mathbb{Z} = \mathbb{Z}_i = \hat{\mathcal{X}} \pm \sqrt{\frac{\sum_{i=1}^n \alpha_i + \beta}{m\alpha_i}} S_i\right) = \frac{m\alpha_i}{4(\sum_{i=1}^n \alpha_i + \beta)},$$

$$\mathbb{P}\left(\mathbb{Z} = \mathbb{Z}_{i+n} = \hat{\mathcal{X}} \pm \sqrt{\frac{\sum_{i=1}^n \alpha_i + \beta}{(1-m)\alpha_i}} S_i\right) = \frac{(1-m)\alpha_i}{4(\sum_{i=1}^n \alpha_i + \beta)},$$

for $i = 1, \cdots, n$. Hence we have a total of $4n + 1$ sigma points. This discrete distribution satisfies $\mathbb{E}(\mathbb{Z}) = \mu$ and $\mathbb{E}\left[(\mathbb{Z} - \hat{\mathcal{X}})(\mathbb{Z} - \hat{\mathcal{X}})^T\right] = \Sigma^{\mathcal{X}\mathcal{X}}$, with the sum of probabilities as unity.

Lemma 2 *The mean and covariance of the discrete random vector $\mathbb{Z}$ with weight probabilities w_i are $\mathbb{E}\left[\mathbb{Z}\right] = \sum_i w_i \mathbb{Z}_i = \hat{\mathcal{X}}$ and $\mathbb{E}\left[(\mathbb{Z} - \mu)(\mathbb{Z} - \mu)^T\right] = \Sigma^{\mathcal{X}\mathcal{X}}$.*

Proof: For $4n + 1$ sigma points and taking $\Psi = \sum_{i=1}^{n} \alpha_i + \beta$, the expected value of $\mathbb{Z}$ can be calculated as

$$
\mathbb{E}\left[\mathbb{Z}\right] = \sum_{i=1}^{4n+1} w_i \mathbb{Z}_i = \mu \left(1 - \frac{\sum_{i=1}^{n} \alpha_i}{2\Psi}\right) + \sum_{i=1}^{n} \left(\frac{m\alpha_i}{4\Psi}\right)
$$

$$
\times \left(\hat{\mathcal{X}} + \sqrt{\frac{\Psi}{m\alpha_i}} S_i\right) + \sum_{i=1}^{n} \left(\frac{m\alpha_i}{4\Psi}\right) \left(\hat{\mathcal{X}} - \sqrt{\frac{\Psi}{m\alpha_i}} S_i\right)
$$

$$
+ \sum_{i=1}^{n} \left(\frac{(1-m)\alpha_i}{4\Psi}\right) \times \left(\hat{\mathcal{X}} + \sqrt{\frac{\Psi}{(1-m)\alpha_i}} S_i\right)
$$

$$
+ \sum_{i=1}^{n} \left(\frac{(1-m)\alpha_i}{4\Psi}\right) \times \left(\hat{\mathcal{X}} - \sqrt{\frac{\Psi}{(1-m)\alpha_i}} S_i\right)
$$

$$
= \left(1 - \frac{\sum_{i=1}^{n} \alpha_i}{2\Psi}\right) \mu + 2 \sum_{i=1}^{n} \left(\frac{m\alpha_i}{4\Psi}\right) \mu + 2 \sum_{i=1}^{n} \left(\frac{\alpha_i \mu}{4\Psi}\right)
$$

$$
- 2 \sum_{i=1}^{n} \left(\frac{m\alpha_i}{4\Psi}\right) \hat{\mathcal{X}} = \hat{\mathcal{X}}.
$$

Similarly, for the covariance matrix, we have

$$
\mathbb{E}\left[(\mathbb{Z} - \hat{\mathcal{X}})(\mathbb{Z} - \hat{\mathcal{X}})^T\right] = \sum_{i=1}^{4n+1} w_i (\mathbb{Z}_i - \hat{\mathcal{X}})(\mathbb{Z}_i - \hat{\mathcal{X}})^T
$$

$$
= \sum_{i=1}^{n} \left(\frac{m\alpha_i}{4\Psi}\right) \Theta\Theta^T + \sum_{i=1}^{n} \left(\frac{m\alpha_i}{4\Psi}\right) \tilde{\Theta}\tilde{\Theta}^T + \sum_{i=1}^{n} \left(\frac{(1-m)\alpha_i}{4\Psi}\right)
$$

$$
\times \Lambda\Lambda^T + \sum_{i=1}^{n} \left(\frac{(1-m)\alpha_i}{4\Psi}\right) \tilde{\Lambda}\tilde{\Lambda}^T
$$

$$
= \sum_{i=1}^{n} \left(\frac{m\alpha_i}{4\Psi}\right) \left(\frac{\Psi}{m\alpha_i}\right) S_i S_i^T + \sum_{i=1}^{n} \left(\frac{m\alpha_i}{4\Psi}\right) \left(\frac{\Psi}{m\alpha_i}\right) S_i S_i^T
$$

$$
+ \sum_{i=1}^{n} \left(\frac{(1-m)\alpha_i}{4\Psi}\right) \left(\frac{\Psi}{(1-m)\alpha_i}\right) S_i S_i^T + \sum_{i=1}^{n} \frac{(1-m)\alpha_i}{4\Psi}
$$

$$
\times \left(\frac{\Psi}{(1-m)\alpha_i}\right) S_i S_i^T = \sum_{i=1}^{n} S_i S_i^T = \Sigma^{\mathcal{X}\mathcal{X}}. \qquad \blacksquare
$$

Here, $\Theta = \sqrt{\frac{\Psi}{m\alpha_i}} S_i$, $\tilde{\Theta} = -\sqrt{\frac{\Psi}{m\alpha_i}} S_i$, $\Lambda = \sqrt{\frac{\Psi}{(1-m)\alpha_i}} S_i$ and $\tilde{\Lambda} = -\sqrt{\frac{\Psi}{(1-m)\alpha_i}} S_i$

.

The next lemma establishes the condition on parameter β under which the maximum probability weight corresponds to the mean.

Lemma 3 $\mathbb{P}(\mathbb{Z} = \hat{\mathcal{X}}) > \mathbb{P}(\mathbb{Z} = \mathbb{Z}_j)$, *for* $j = 1, \cdots, 4n$, *if the following condition holds:* $\beta \geq \frac{1}{4} \max{(m\alpha_i)} - \frac{1}{2} \sum_{i=1}^{n} \alpha_i$.

Proof: From definition, $\mathbb{P}(\mathbb{Z} = \hat{\mathcal{X}}) > \mathbb{P}(\mathbb{Z} = \mathbb{Z}_j)$, $j = 1, \cdots, 4n$ if

$$1 - \frac{\sum_{i=1}^{n} \alpha_i}{2\Psi} > \max \left(\frac{m\alpha_i}{4\Psi}, \frac{(1-m)\alpha_i}{4\Psi} \right), \text{ i.e.,}$$

$$\frac{\sum_{i=1}^{n} \alpha_i + 2\beta}{2\Psi} > \frac{1}{4\Psi} \max{(m\alpha_i)}, \text{ i.e.,}$$

$$\sum_{i=1}^{n} \alpha_i + 2\beta > \frac{1}{2} \max{(m\alpha_i)},$$

which leads to the required result. ∎

The sigma point generation algorithm has two parameters: β and m. Heuristics for choosing these parameters are explained below.

- The choice of β helps in establishing the main idea of the new sigma point transformation that the points nearer to the mean shall have more probability of occurrence, i.e., weights (w_i). From Lemma 2, we can write $\beta > \beta_{min}$ where $\beta_{min} = \frac{1}{4}\max(m\alpha_i) - \frac{1}{2}\sum_{i=1}^{n} \alpha_i$. If $\beta = \beta_{min}$, out of $4n + 1$ weights, $w_1 = w_i$, where $i = 2, \cdots, 2n + 1$, which is clearly not what we propose. Hence β should be greater than β_{min}. If $\beta \gg \beta_{min}$, most of the probability mass is concentrated at the mean and all the other sigma points have negligible weights. The choice of β needs to ensure that all the sigma points have significant weights, which leads to a suggestion that we choose β to be greater, but not hugely greater than β_{min}.

- The parameter m distributes the probability mass corresponding to S_i to four different support vectors, such that the number of sigma points around the mean is doubled. m being close to 1 makes one set of weights $(2n)$ negligible while $m = 0.5$ makes the two sets $(4n$ points$)$ coincide. Then we will be left with only $2n + 1$ points. For m values greater than 1, one set of points will become imaginary and their probability weights negative, which is not acceptable. Hence we choose $0.5 < m < 1$. A choice $m = 0.6$ seems to be adequate in the examples tried.

It is further worth noting that the definition of α_i does not affect the matching of the first two moments. Thus the algorithm can easily be modified to match a property other than adjusting the probabilities of S_i according to their inner product with the mean.

Apart from the sigma point generation, the actual filtering algorithm follows the same steps as in Section 3.3. Finally, the proposed filter in this section, called the new sigma point filter (NSKF) by its authors, can be easily extended for formulating its square-root algorithm described previously.

3.4.3 MATLAB-based filtering exercises

- Implement the spherical simplex UKF and the NSKF for the system described in Section 3.3.1. Compare the performance of different filters, including the EKF and the three filters discussed in this chapter.

- As a second example, consider a nonlinear state space system used as an illustration in [73] and [128]:

$$\mathcal{X}_{k+1} = \mathcal{X}_k + 5T\mathcal{X}_k(1 - \mathcal{X}_k^2) + \eta_k,$$
$$\mathcal{Y}_k = T\mathcal{X}_k(1 - 0.5\mathcal{X}_k) + v_k,$$

where the noise variables are distributed as: $v_k \sim \mathcal{N}(0, b^2 T)$, $\eta_k \sim \mathcal{N}(0, c^2 T)$. The parameters are given by $T = 0.001$, $b = 0.5$ and $c = 0.11$. Starting with $\Sigma_{0|0}^{\mathcal{X}\mathcal{X}} = 2$ and $\mathcal{X}_0 = -0.8$, carry out 200 simulation runs, each with 400 time-steps for the EKF, the UKF and the NSKF using this system. For the NSKF, use $m = 0.6$ and $\beta = 0.25 \times 0.6 - 0.5 + 0.5 = 0.15$. Note that $\alpha = 1$ for a single dimensional system and it need not be computed at every step.

3.5 Summary

This chapter outlined some of the main ideas behind unscented Kalman filtering, which has now become established as a very useful set of heuristics for real time filtering. The main difference between the UKF and the EKF is the point where linearization is performed; however, both the filters still assume a jointly Gaussian update and share the same risk of divergence if the system dynamics is strongly nonlinear. One of the main advantages of the UKF is that it avoids computing the Jacobians entirely. Apart from the filtering algorithms discussed in this chapter, there are variants of the UKF which aim to match statistical properties in addition to the mean and the variance; these include [166] and [120].

The deterministic sample points, or sigma points in the UKF are computed purely with an objective of matching the leading order terms in the Taylor expansion of the first two moments. It is possible to focus on the optimality of the computation of expectation integrals, and come up with a better set of sample points from an optimal quadrature point of view. This is the basis of the filtering schemes discussed in subsequent chapters.

Chapter 4

Filters based on cubature and quadrature points

4.1	Introduction	65
4.2	Spherical cubature rule of integration	66
4.3	Gauss-Laguerre rule of integration	67
4.4	Cubature Kalman filter	68
4.5	Cubature quadrature Kalman filter	70
	4.5.1 Calculation of cubature quadrature (CQ) points	70
	4.5.2 CQKF algorithm	71
4.6	Square root cubature quadrature Kalman filter	75
4.7	High-degree (odd) cubature quadrature Kalman filter	77
	4.7.1 Approach	77
	4.7.2 High-degree cubature rule	77
	4.7.3 High-degree cubature quadrature rule	79
	4.7.4 Calculation of HDCQ points and weights	80
	4.7.5 Illustrations	80
	4.7.6 High-degree cubature quadrature Kalman filter	86
4.8	Simulation examples	87
	4.8.1 Problem 1	87
	4.8.2 Problem 2	91
4.9	Summary	92

4.1 Introduction

The prior and posterior probability density function (pdf) of states for a nonlinear system are in general non-Gaussian, due to nonlinearities in the process dynamics or in the measurement equation. Under such circumstances, which are commonly observed in practice, estimating the states of a system is a challenge. There are several approaches which have been discussed in Chapter 1 to solve such problems. Let us recall all the assumptions mentioned in Subsection 1.7.2 which includes the assumption mentioned inside the box and the assumption of the prior and the posterior as Gaussian pdfs. As the pdfs are assumed to be Gaussian, the prior and the posterior mean and covariance become extremely important and are required to be determined by filtering algorithm. In other words, the integrals (1.48) to (1.52) need to be evaluated. If we see the structure of the integrals, it is in the form of $\int nonlinear\ function \times Gaussian\ pdf$. For an arbitrary nonlinear function the integral cannot be evaluated analytically. So we have to rely on numerical methods for approximate solution of the integrals.

The reader may think that the integrals can be evaluated with the sum over many points on real space. But the method is impractical as the integral limit ranges from $+\infty$ to $-\infty$ and the evaluation at each time step becomes so time consuming that it is impossible to implement it for a real time estimation problem with commonly available hardware. Instead, we need a few points which are wisely chosen so that summation over such points provides a sufficiently close approximation of an integral. In this chapter, we shall discuss the evaluation of such an integral with the help of cubature quadrature support points and accordingly we shall discuss the filtering algorithm based on it. We shall plan to proceed in line with how the subject has been historically developed.

4.2   Spherical cubature rule of integration

To compute the integrals (1.48) to (1.52) of dimension n of a nonlinear function over a Gaussian pdf, at first, a spherical radial transformation is applied, where the variable from the Cartesian vector $x \in \mathbb{R}^n$ is changed to a radius r and the direction vector Z. The straightforward derivation is given here in the form of a theorem.

Theorem 1 *For an arbitrary function* $\phi(x)$, $x \in \mathbb{R}^n$, *the integral*

$$I(\phi) = \frac{1}{\sqrt{|\Sigma|(2\pi)^n}} \int_{\mathbb{R}^n} \phi(x) e^{-\frac{1}{2}(x-\mu)^T \Sigma^{-1}(x-\mu)} dx$$

can be expressed in a spherical coordinate system as

$$I(\phi) = \frac{1}{\sqrt{(2\pi)^n}} \int_{r=0}^{\infty} \int_{\mathbb{U}_n} [\phi(CrZ + \mu) ds(Z)] r^{n-1} e^{r^2/2} dr, \qquad (4.1)$$

where $x = CrZ + \mu$, C *is the Cholesky decomposition of* Σ, $|Z| = 1$. $ds(.)$ *is an area element of the surface of unit hyper-sphere* $\mathbb{U}_n$.

Proof 1 *Let us transform the integral* $I(\phi)$ *to a spherical coordinate system [49]. Let* $x = CY + \mu$, $Y \in \mathbb{R}^n$, *where* $\Sigma = CC^T$; *i.e.,* C *is the Cholesky decomposition of* Σ. *Then* $(x - \mu)^T \Sigma^{-1}(x - \mu) = Y^T C^T C^{-T} C^{-1} CY = Y^T Y$ *and* $dx = |C|dY = \sqrt{|\Sigma|}dY$. *So the integral of interest becomes*

$$I(\phi) = \frac{1}{\sqrt{(2\pi)^n}} \int_{\mathbb{R}^n} \phi(CY + \mu) e^{-\frac{1}{2}Y^T Y} dY. \qquad (4.2)$$

Now, let $Y = rZ$, with $|Z| = \sqrt{Z^T Z} = 1$, $Y^T Y = Z^T rr Z = r^2$. *The elementary volume of a hyper-sphere at n dimensional space is* $dY = r^{n-1} dr ds(Z)$, *where* $ds(.)$ *is the area element on* $\mathbb{U}_n$. $\mathbb{U}_n$ *is the surface of a hyper-sphere defined by* $\mathbb{U}_n = \{Z \in \mathbb{R}^n | Z Z^T = 1\}$; $r \in [0, \infty)$. *Hence*

$$
\begin{aligned}
I(\phi) &= \frac{1}{\sqrt{(2\pi)^n}} \int_{r=0}^{\infty} \int_{\mathbb{U}_n} \phi(CrZ + \mu) e^{-r^2/2} r^{n-1} dr ds(Z) \\
&= \frac{1}{\sqrt{(2\pi)^n}} \int_{r=0}^{\infty} \int_{\mathbb{U}_n} [\phi(CrZ + \mu) ds(Z)] r^{n-1} e^{-r^2/2} dr.
\end{aligned}
\tag{4.3}
$$

∎

From the above expression we see that an n dimensional volume integral is split into a surface and a line integral. Now to compute the integration $I(\phi)$ as described above, first we need to compute

$$
\int_{\mathbb{U}_n} \phi(CrZ + \mu) \, ds(Z).
\tag{4.4}
$$

The spherical integral (4.4) can be approximately calculated by the third degree fully symmetric spherical cubature rule [107, 52]. If we consider zero mean unity variance (i.e., $\mu = 0$ and $\Sigma = I$), Eq. (4.4) can be approximated as [5]

$$
\int_{\mathbb{U}_n} \phi(rZ) ds(Z) \approx \frac{2\sqrt{\pi^n}}{2n\Gamma(n/2)} \sum_{i=1}^{2n} \phi[ru_i],
\tag{4.5}
$$

where Γ is the Gamma function and $[u_i]$ ($i = 1, 2, \cdots, 2n$) are the cubature points located at the intersections of the unit hypersphere and its axes for a fully symmetric cubature rule. For example, in a single dimension, the two cubature points will be on $+1$ and -1. For two dimensions, the four cubature points will be on $(+1, 0)$, $(-1, 0)$, $(0, +1)$ and $(0, -1)$. For a Gaussian distribution with non zero mean μ and non unity covariance C, the cubature points will be located at $(C[u_i] + \mu)$. All the points have the same weight for the fully symmetric spherical cubature rule. Also the cubature rule is exact for the polynomial function $\phi(x)$ up to degree three [5].

4.3 Gauss-Laguerre rule of integration

Any integral of a function $\phi(\lambda)$ in the form of

$$
\int_{\lambda=0}^{\infty} \phi(\lambda) \lambda^{\alpha} e^{-\lambda} d\lambda
\tag{4.6}
$$

(where α is an integer) can be approximately evaluated with the weighted sum of quadrature points. The quadrature points are the roots of the n order of the Chebyshev-Laguerre polynomial [96, 69] given by

$$L_{n'}^{\alpha}(\lambda) = (-1)^{n'} \lambda^{-\alpha} e^{\lambda} \frac{d^{n'}}{d\lambda^{n'}} \lambda^{\alpha+n'} e^{-\lambda} = 0. \tag{4.7}$$

Let the quadrature points be $\lambda_{i'}$. The weights can be determined as

$$A_{i'} = \frac{n'! \Gamma(\alpha + n' + 1)}{\lambda_{i'} [\dot{L}_{n'}^{\alpha}(\lambda_{i'})]^2}. \tag{4.8}$$

The integer n' is user choice and accuracy of the summation depends on it. To summarize, the integral (4.6) can be written approximately with n' number of support points and weights using the Gauss-Laguerre quadrature rule as

$$\int_{\lambda=0}^{\infty} \phi(\lambda) \lambda^{\alpha} e^{-\lambda} d\lambda \approx \sum_{i'=1}^{n'} A_{i'} \phi(\lambda_{i'}). \tag{4.9}$$

4.4 Cubature Kalman filter

The filter was developed by Arasaratnam and Heykin [5] in 2009. In the cubature Kalman filer (CKF), the surface integral (4.4) is approximated with the third order spherical cubature rule given by expression (4.5) and the integral Eq. (4.6) is approximated with $n' = 1$ for which the polynomial becomes

$$L_1^{\alpha}(\lambda) = -\lambda^{-\alpha} e^{\lambda} \frac{d}{d\lambda} \lambda^{\alpha+1} e^{-\lambda} = 0,$$

or,

$$\lambda = \alpha + 1.$$

Combining Eqs. (4.5) and (4.1) we get

$$I(\phi) = \frac{1}{\sqrt{(2\pi)^n}} \times \frac{2\sqrt{\pi^n}}{2n\Gamma(n/2)} \int_{r=0}^{\infty} \left(\sum_{i=1}^{2n} \phi[ru_i]\right) r^{n-1} e^{-r^2/2} dr. \tag{4.10}$$

Now to integrate the underlined term, the Gauss-Laguerre rule of integration described above can be used. By substituting $t = r^2/2$, the above equation becomes

$$I(\phi) = \frac{1}{2n\Gamma(n/2)} \int_{t=0}^{\infty} \left(\sum_{i=1}^{2n} \phi[\sqrt{2t}u_i]\right) t^{(n/2-1)} e^{-t} dt. \tag{4.11}$$

Now the integration $\int_{t=0}^{\infty} \phi(t) t^{(n/2-1)} e^{-t} dt$ is approximated using the points obtained from the first order Gauss-Laguerre polynomial; i.e., $n' = 1$ and the expression of α becomes $\alpha = n/2 - 1$. Eq. (4.11) can further be expressed as

$$I(\phi) = \frac{1}{2n\Gamma(n/2)} \times [\sum_{i=1}^{2n} A_1 \phi(\sqrt{2(n/2)})[u_i]]. \tag{4.12}$$

For $n' = 1$, i.e., the first order approximation of line integral using the Gauss-Laguerre polynomial, the weight term becomes

$$A_1 = \frac{\Gamma(n/2 - 1)}{n/2[\dot{L}_1(\lambda_1)]^2} = (n/2 - 1)!. \tag{4.13}$$

Substituting the value of A_1 in the Eq. (4.12), we receive

$$I(\phi) = \frac{1}{2n\Gamma(n/2)} \times (n/2 - 1)! \times [\sum_{i=1}^{2n} \phi(\sqrt{2(n/2)})[u_i]], \tag{4.14}$$

or,

$$I(\phi) = \frac{1}{2n} \times [\sum_{i=1}^{2n} \phi(\sqrt{n})[u_i]]. \tag{4.15}$$

The cubature Kalman filter which assumes $n' = 1$, and the third order spherical cubature rule, requires $2n$ number of support points (n is the order of a system) with the value $\sqrt{n}[u_i]$, and weights $1/2n$. Note that $[u_i]$ is the unity value in i^{th} coordinate.

Illustration:

(a) Single dimensional case ($n = 1$): Two cubature points are $\{1 \quad -1\}$ and their corresponding weights are $\{0.5 \quad 0.5\}$.

(b) Two dimensional case ($n = 2$): In two dimensional state space the four cubature points will be at $\left\{ \begin{array}{cccc} 1.414 & 0 & -1.414 & 0 \\ 0 & 1.414 & 0 & -1.414 \end{array} \right\}$, and all of them have the same weights 0.25.

(c) Three dimensional system ($n = 3$): In three dimensional state space the six cubature points are at

$$\left\{ \begin{array}{cccccc} 1.732 & 0 & -1.732 & 0 & 0 & 0 \\ 0 & 0 & 0 & 0 & 1.732 & -1.732 \\ 0 & 1.732 & 0 & -1.732 & 0 & 0 \end{array} \right\},$$

and all of them have the same weights $1/6$.

We understand, at this point, the readers are eager to know how those points will be utilized to estimate the states of a system. Once the points and weights are generated, the estimation algorithm is easy to implement. We would like to suggest that the reader be patient until we describe all the cubature methods which will be used to generate points and weights. Once the task is complete, we shall navigate to the implementable algorithm section.

4.5   Cubature quadrature Kalman filter

The CKF is further generalized by Bhaumik and Swati [17] [16] and the filter was named the cubature quadrature Kalman filter (CQKF). The CQKF uses the third order spherical cubature rule for integrating the surface integral and any arbitrary order n', Gauss-Laguerre quadrature points. As a result the total $2nn'$ points and associated weights need to be calculated. These points are named cubature quadrature (CQ) points.

Let us recall Eq. (4.11), and let us evaluate the line integral with the n' order Gauss-Laguerre polynomial. In such a situation, Eq. (4.11) can be further written as

$$I(\phi) = \frac{1}{2n\Gamma(n/2)} \times [\sum_{i=1}^{2n} \sum_{i'=1}^{n'} A_{i'} \alpha(\sqrt{2\lambda_{i'}})[u_i]]. \tag{4.16}$$

4.5.1   Calculation of cubature quadrature (CQ) points

The steps for calculating support points and associated weights are as follows:

- Find the cubature points $[u_i]_{(i=1,2,\cdots,n)}$, located at the intersection of the unit hyper-sphere and its axes.

- Solve the n' order Chebyshev-Laguerre polynomial (Eq. (4.7)) for $\alpha = (n/2 - 1)$ to obtain the quadrature points $(\lambda_{i'})$.

$$L_{n'}^{\alpha}(\lambda) = \lambda^{n'} - \frac{n'}{1!}(n'+\alpha)\lambda^{n'-1} + \frac{n'(n'-1)}{2!}(n'+\alpha)(n'+\alpha-1)\lambda^{n'-2} - \cdots = 0$$

- Find the cubature quadrature (CQ) points as $\xi_j = \sqrt{2\lambda_{i'}}[u_i]$ and their corresponding weights as

$$w_j = \frac{1}{2n\Gamma(n/2)}(A_{i'}) = \frac{1}{2n\Gamma(n/2)} \frac{n'!\Gamma(\alpha+n'+1)}{\lambda_{i'}[\dot{L}_{n'}^{\alpha}(\lambda_{i'})]^2},$$

for $i = 1, 2, \cdots, 2n$, $i' = 1, 2, \cdots, n'$ and $j = 1, 2, \cdots, 2nn'$.

It should be noted that the sum of all weights of all the support points is unity. So the CQ points and the associated weights represent the probability mass function.

Cubature quadrature points are the generalization of cubature points. If we assume first order Gauss-Laguerre approximation, the cubature quadrature points merge with cubature points as proposed by Araratnam and Heykin [5]. As the value of n' increases, the calculation of the integral becomes more accurate. MATLAB code for generating CQ points and corresponding weights is provided in Listing 4.1. The values of CQ points and corresponding weights for different n' value are listed in Table 4.1. The CQKF points and weights for a second order system have been plotted in Figure 4.1

```
1 % Program to generate cubature quadrature points
2 % n: state dimension
3 % n_prime: Order of Gauss-Laguerre quadrature
4 % cq_pts: CQ points (2*n*n_prime) x n
5 % cq_wts: weights (2*n*n_prime) x 1
6 function [cq_pts, cq_wts] = cubquad_points(n, n_prime)
7 alpha = 0.5*n - 1;
8 syms x;
9 CL_poly=laguerreL(n_prime,alpha,x); %Chebyshev - Laguerre    poly
10 CL_poly_d=diff(CL_poly);%Differential of Chebyshev-Laguerre
        polynomial
11 q_pts=real(double(roots(flip(coeffs(CL_poly)))));
12 aa=(factorial(n_prime)*gamma(alpha+n_prime+ 1)) / ...
13          (2*n*gamma(n/ 2));
14 dd = double(subs(CL_poly_d, q_pts)) .^ 2;
15 q_wts = aa ./ (q_pts .* dd);
16 q_pts = sqrt(2 * q_pts);
17
18 for j = 1:n_prime
19   low = (2*n*(j - 1))+1;
20   high = 2*n*j;
21   cq_pts(low:high,:)=q_pts(j)*[diag(ones(1,n));-diag(ones(1, n))
        ];
22   cq_wts(low:high,1)=repmat(q_wts(j),[2*n,1]);
23 end
24
25 cq_wts = cq_wts ./ sum(cq_wts);
26
27 end
```

Listing 4.1: Generation of CQ points and weights.

4.5.2   CQKF algorithm

A filtering algorithm can easily be constructed with the help of the CQ rule of integration described above. The arbitrary vector x considered above now becomes our state vector $\mathcal{X}$ whose pdfs need to be approximated. The algorithm of the cubature quadrature Kalman filter (CQKF) is provide in Algorithm 8. Several observations on the above algorithm are in order.

- Compared to the extended Kalman filter (EKF), the filters in the cubature family are derivative free; i.e., to implement them, neither a Jacobian nor Hessian matrix needs to be calculated. This may be considered as an added advantage from the computational point of view.

- The accuracy of the filter depends on the order of the Gauss-Laguerre quadrature. The higher the order, the more accurate the estimator would be.

TABLE 4.1: CQ points and weights for different n and n'

System dimension	Order of radial rule (n')	Points	Weights
1	1	(+1.0), (-1.0)	0.5, 0.5
	2	(2.3344), (-2.3344), (0.7420), (-0.7420)	2.3344, -2.3344, 0.7420, -0.7420
	3	(3.3243), (-3.3243), (1.8892), (-1.8892), (0.6167), (-0.6167)	0.0026, 0.0026, 0.0886, 0.0886, 0.4088, 0.4088
2	1	(1.4142,0), (-1.4142,0), (0, 1.4142), (0,-1.4142)	0.25, 0.25, 0.25, 0.25,
	2	(2.6131,0), (-2.6131,0), (0, 2.6131), (0, -2.6131), (1.0842,0), (-1.0842,0), (0, 1.0842), (0, -1.0842)	0.0366, 0.0366, 0.0366, 0.0366, 0.2134, 0.2134, 0.2134, 0.2134
	3	(3.5468,0), (-3.5468,0), (0,3.5468), (0,-3.5468), (2.1421,0), (-2.1421,0), (0,2.1421), (0,-2.1421), (0.9119,0), (-0.9119,0), (0,0.9119), (0,-0.9119)	0.0026, 0.0026, 0.0026, 0.0026, 0.0696, 0.0696, 0.0696, 0.0696, 0.1778, 0.1778, 0.1778, 0.1778
3	1	(1.7321,0,0), (-1.7321,0,0), (0,1.7321,0), (0,-1.7321,0), (0,0,1.7321), (0,0,-1.7321),	0.1667, 0.1667, 0.1667, 0.1667, 0.1667, 0.1667
	2	(2.857,0,0), (-2.857,0,0), (0, 2.857,0), (0, -2.857,0), (0, 0, 2.857), (0,0, -2.857) (1.3556,0,0), (-1.3556,0,0), (0, 1.3556,0), (0, -1.3556,0), (0, 0, 1.3556), (0,0, -1.3556)	0.0306, 0.0306, 0.0306, 0.0306, 0.0306, 0.0306, 0.136, 0.136, 0.136, 0.136, 0.136, 0.136
	3	(3.7504,0,0), (-3.7504,0,0), (0,3.7504,0), (0,-3.7504,0), (0,0,3.7504), (0,0,-3.7504) (2.3668,0,0), (-2.3668,0,0), (0,2.3668,0), (0,-2.3668,0), (0,0,2.3668), (0,0,-2.3668), (1.1544,0,0), (-1.1544,0,0), (0,1.1544,0), (0,-1.1544,0), (0,0,1.1544), (0,0,-1.1544)	0.0026, 0.0026, 0.0026, 0.0026, 0.0026, 0.0026, 0.0574, 0.0574, 0.0574, 0.0574, 0.0574, 0.0574, 0.1067, 0.1067, 0.1067, 0.1067, 0.1067, 0.1067

Algorithm 8 Cubature quadrature Kalman filter

$$[\hat{\mathcal{X}}_{k|k}, \Sigma_{k|k}] = \text{CQKF}[\mathcal{Y}_k]$$

- Initialize the filter with $\hat{\mathcal{X}}_{0|0}$ and $\Sigma_{0|0}$.

- For $j = 1 : 2nn'$
  - Calculate CQ points ξ_j.
  - Calculate w_j.

- End For

- For $k = 1 : N$

  Prediction step

  - Cholesky decomposition of posterior error covariance, $\Sigma_{k|k} = S_{k|k}S_{k|k}^T$.
  - Spread the CQ points, $\chi_{j,k|k} = S_{k|k}\xi_j + \hat{\mathcal{X}}_{k|k}$.
  - Compute the prior mean, $\hat{\mathcal{X}}_{k+1|k} = \sum_{j=1}^{2nn'} w_j \chi_{j,k+1|k}$.
  - The prior covariance,
  $\Sigma_{k+1|k} = \sum_{j=1}^{2nn'} w_j[\chi_{j,k+1|k} - \hat{\mathcal{X}}_{k+1|k}][\chi_{j,k+1|k} - \hat{\mathcal{X}}_{k+1|k}]^T + Q_k$.

  Update step

  - Cholesky decomposition of prior error covariance, $\Sigma_{k+1|k} = S_{k+1|k}S_{k+1|k}^T$.
  -Spread CQ points, $\chi_{j,k+1|k} = S_{k+1|k}\xi_j + \hat{\mathcal{X}}_{k+1|k}$.
  - Predicted measurements at CQ points, $Y_{j,k+1|k} = \gamma(\chi_{j,k+1|k})$
  - Predicted measurement, $\hat{\mathcal{Y}}_{k+1} = \sum_{j=1}^{2nn'} w_j Y_{j,k+1|k}$.
  - Measurement error covariance,
  $\Sigma_{k+1|k}^{yy} = \sum_{j=1}^{2nn'} w_j[Y_{j,k+1|k} - \hat{\mathcal{Y}}_{k+1}][Y_{j,k+1|k} - \hat{\mathcal{Y}}_{k+1}]^T + R_k$.
  - Cross covariance,
  $\Sigma_{k+1|k}^{xy} = \sum_{j=1}^{2nn'} w_j[\chi_{j,k+1|k} - \hat{\mathcal{X}}_{k+1|k}][Y_{j,k+1|k} - \hat{\mathcal{Y}}_{k+1}]^T$.
  - Kalman gain, $K_{k+1} = \Sigma_{k+1|k}^{xy}(\Sigma_{k+1|k}^{yy})^{-1}$.
  -Posterior state estimate, $\hat{\mathcal{X}}_{k+1|k+1} = \hat{\mathcal{X}}_{k+1|k} + K_{k+1}(\mathcal{Y}_{k+1} - \hat{\mathcal{Y}}_{k+1})$.
  - Posterior error covariance
  $\Sigma_{k+1|k+1} = \Sigma_{k+1|k} - K_{k+1}\Sigma_{k+1|k}^{yy}K_{k+1}^T$.

- End For

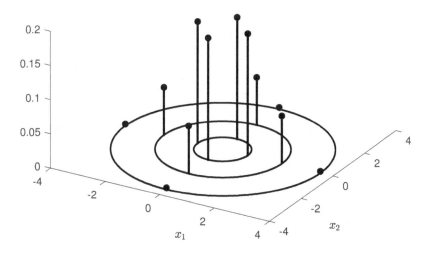

FIGURE 4.1: Plot of CQ points and weights for a 2D system.

- As discussed earlier, the CQKF is computationally efficient compared to the PF and grid based filter because instead of generating large number of particles or grid points in the state space a few support points are required to be generated, updated and predicted. The method is also computationally more efficient than the Gauss-Hermite filter (GHF) because to implement the n'^{th} order Gauss-Hermite filter for an n dimensional system, n'^n number of quadrature points are necessary whereas to implement the n'^{th} order CQKF (CQKF-n') $2nn'$ number of support points are necessary. However, the computational cost of CQKF is slightly higher than EKF and CKF.

- The cubature quadrature points as well as their weights can be calculated and stored off line.

- For the unscented or sigma point Kalman filter, discussed in Section 3.3, the sigma point, located at the center is the most significant and carries the highest weight, whereas in CQKF there is no cubature-quadrature point at the origin. It is concluded in [5] that the cubature approach is "more principled in mathematical terms" than the sigma point approach. The same argument is true for the cubature quadrature Kalman filter.

4.6 Square root cubature quadrature Kalman filter

We see that during implementation of the CQKF and CKF algorithm, the Cholesky decomposition is necessary to be performed at each step. However, due to limited word length in processing software and accumulated round off error associated with processing software, the Cholesky decomposition sometimes leads to negative definite covariance matrix, causing the filter to stop. The unscented Kalman filter (UKF) [83] and the Gauss-Hermite filter (GHF) [74] also suffer from a similar type of hitch which has been rectified by Merwe et al. [170], and Arasaratnam et al. [4], by proposing square-root formulation.

In place of the Cholesky decomposition, Merwe and others [170] proposed a square root technique which has the added benefit of numerical stability and guaranteed positive semi-definiteness of the error covariance matrix. In [4], Arasaratnam and Heykin proposed a square root version of the Gauss-Hermite filter algorithm which uses the Gauss-Hermite quadrature method to evaluate intractable integrals. In this section, we describe the square root version of CQKF (SR-CQKF) [15] based on orthogonal-triangular decomposition, popularly known as QR decomposition. Contrary to ordinary CQKF, where Cholesky decomposition, computationally the most expensive operation, has to be performed at each step, the SR-CQKF propagates the state estimate and its square-root covariance. The square-root version of UKF is now being applied to practical problems [149, 165]. Square root CQKF has also found application to satellite attitude determination [93].

The error covariance matrix, Σ can be factorized as

$$\Sigma = AA^T, \tag{4.17}$$

where $\Sigma \in \mathbb{R}^{n \times n}$, and $A \in \mathbb{R}^{n \times m}$ with $m \geq n$. Due to increased dimension of A, Arasaratnam and Haykin [4] called it a "fat" matrix. The QR decomposition of matrix A^T is given by, $A^T = QR$ where $Q \in \mathbb{R}^{m \times m}$ is orthogonal, and $R \in \mathbb{R}^{m \times n}$ is upper triangular (please do not mix Q and R in this section with noise covariances). With the QR decomposition described above, the error covariance matrix can be written as

$$\Sigma = AA^T = R^T Q^T QR = \tilde{R}^T \tilde{R} = SS^T. \tag{4.18}$$

$\tilde{R}$ is the upper triangular part of the R matrix, $\tilde{R} \in \mathbb{R}^{n \times n}$, where $\tilde{R}^T = S$.

The detailed algorithm of the square-root cubature-quadrature Kalman filter is provided in Algorithm 9. We use the notation $qr\{\}$ to denote the QR decomposition of a matrix and $uptri\{\}$ to select the upper triangular part of a matrix. The equations mentioned in the algorithm can be proved. For detailed steps readers are referred to [5].

Note: The practitioners, who want to implement the algorithm in the MATLAB software environment, should note that the command

---

Algorithm 9 Square-root cubature quadrature Kalman filter

$$[\hat{\mathcal{X}}_{k|k}, \Sigma_{k|k}] = \text{SRCQKF}[\mathcal{Y}_k]$$

- Initialize the filter with mean $\hat{\mathcal{X}}_{0|0}$ and $\Sigma_{0|0}$.

- Cholesky factorize $\Sigma_{0|0} = S_{0|0}S_{0|0}^T$

- For $j = 1 : 2nn'$
  - Calculate CQ points ξ_j.
  - Calculate w_j.

- End For

- For $k = 1 : N$

  Prediction step

  - Spread the CQ points, $\chi_{j,k|k} = S_{k|k}\xi_j + \hat{\mathcal{X}}_{k|k}$.
  - Compute the prior mean, $\hat{\mathcal{X}}_{k+1|k} = \sum_{j=1}^{2nn'} w_j \chi_{j,k+1|k}$.
  - The prior covariance,
  $\Sigma_{k+1|k} = \sum_{j=1}^{2nn'} w_j[\chi_{j,k+1|k} - \hat{\mathcal{X}}_{k+1|k}][\chi_{j,k+1|k} - \hat{\mathcal{X}}_{k+1|k}]^T + Q_k$.

  - Sqrt. weight matrix, $W = \begin{bmatrix} \sqrt{w_1} & \cdots & 0 \\ \vdots & \ddots & \vdots \\ 0 & \cdots & \sqrt{w_{2nn'}} \end{bmatrix}$.

  - Calculate
  $\chi_{k+1|k}^* = [\chi_{1,k+1|k} - \hat{\mathcal{X}}_{k+1|k} \ \chi_{2,k+1|k} - \hat{\mathcal{X}}_{k+1|k} \ \cdots \ \chi_{2nn',k+1|k} - \hat{\mathcal{X}}_{k+1|k}]W$.
  - Perform QR decomposition, $\Re_{k+1|k} = qr[\chi_{k+1|k}^* \ \sqrt{Q_k}]$.
  - Sqrt. of prior error covariance, $S_{k+1|k} = uptri\{\Re_{k+1|k}\}$.

  Update step

  - Spread CQ points, $\chi_{j,k+1|k} = S_{k+1|k}\xi_j + \hat{\mathcal{X}}_{k+1|k}$.
  - Predicted measurements at CQ points, $Y_{j,k+1|k} = \gamma(\chi_{j,k+1|k})$.
  - Predicted measurement, $\hat{\mathcal{Y}}_{k+1} = \sum_{j=1}^{2nn'} w_j Y_{j,k+1|k}$.
  - Sqrt. weighted measurement matrix,
  $Y_{k+1|k}^* = [Y_{1,k+1|k} - \hat{\mathcal{Y}}_{k+1|k} \ Y_{2,k+1|k} - \hat{\mathcal{Y}}_{k+1|k} \ \cdots \ Y_{2nn',k+1|k} - \hat{\mathcal{Y}}_{k+1|k}]W$.
  - Perform QR decomposition, $\Re_{\mathcal{Y}_{k+1}\mathcal{Y}_{k+1}} = qr[Y_{k+1|k}^* \ \sqrt{R_k}]$.
  - Sqrt. of innovation covariance, $S_{\mathcal{Y}_{k+1}\mathcal{Y}_{k+1}} = uptri\{\Re_{\mathcal{Y}_{k+1}\mathcal{Y}_{k+1}}\}$.
  - Cross covariance matrix, $\Sigma_{k+1|k}^{\mathcal{X}\mathcal{Y}} = \chi_{k+1|k}^* Y_{k+1|k}^*$.
  - Kalman gain, $K = (\Sigma_{k+1}^{\mathcal{X}\mathcal{Y}}/S_{\mathcal{Y}_{k+1}\mathcal{Y}_{k+1}}^T)/S_{\mathcal{Y}_{k+1}\mathcal{Y}_{k+1}}$.
  - Posterior estimate, $\hat{\mathcal{X}}_{k+1|k+1} = \hat{\mathcal{X}}_{k+1|k} + K(\mathcal{Y}_{k+1} - \hat{\mathcal{Y}}_{k+1})$.
  - QR decomposition, $\Re_{k+1|k+1} = qr[\chi_{k+1|k}^* - KY_{k+1|k}^* \ \sqrt{Q_k} \ K\sqrt{R_k}]$.
  - Sqrt. of posterior error covariance,
  $S_{k+1k+1} = uptri\{\Re_{k+1|k+1}\}$.

- End For

$chol(\Sigma)$ returns the matrix S^T (where $\Sigma = SS^T$) rather than S. Accordingly the expressions for the next steps need to be evaluated.

4.7 High-degree (odd) cubature quadrature Kalman filter

4.7.1 Approach

Arasaratnam and Heykin in [5] stated their experience that high-degree cubature rules yield "no improvement or make the performance worse". However, the statement was questioned by Bin Jia et al. [81] who argued that high-degree CKFs are necessary to obtain more accurate results and formulated a high-degree cubature Kalman filter (HDCKF). They used the CKF like approach where the said integral is split into a surface and radial integral. The surface integral over a hyper-sphere is calculated using the spherical cubature rule of arbitrary accuracy [81, 50]. The arbitrary accuracy is achieved by using any odd high-degree spherical cubature rule to solve the surface integral. The radial integral was solved with the moment matching method [81]. The proposed estimator works more accurately compared to CKF.

Singh et al. [142] modified the above approach where any arbitrary order Gauss-Laguerre quadrature method [96, 69] is used to solve the radial integral, and the high-degree (odd) cubature rule [81, 50] is used to evaluate the surface integral. As a result, a new algorithm has been evolved which is named the high-degree cubature quadrature Kalman filter (HDCQKF). HDCQKF is more accurate compared to the HDCKF, CKF, and CQKF and the heuristics associated with the moment matching method in HDCKF [81] could be avoided. The two parameters of the HDCQKF, namely the degree of cubature approximation and the order of quadrature approximation may be varied to control the estimation accuracy versus computational efficiency. Moreover, all the filters which belong to the cubature family would be a special case of the HDCQKF. In this section, we shall discuss the procedure to generate the points and the weights for HDCQKF. Once the points and the weights are generated the filtering algorithm mentioned above could be implemented to obtain the state estimate. We request that the reader recall Theorem 1. The concept of HDCQKF is illustrated with the help of Theorem 1 and the following three theorems described in the next subsection.

4.7.2 High-degree cubature rule

Recalling Theorem 1, it is clear that our first task is to calculate the integral, $\int_{U_n} \phi(CrZ + \mu) d\sigma(Z)$ over the surface of a hyper-sphere of dimension

n. Without loss of generality, if we assume zero mean and unity covariance the above integral becomes $\int_{U_n} \phi(rZ) d\sigma(Z)$.

To solve a multidimensional integral numerically, various types of cubature rule exist in literature [156, 32]. In HDCQKF, a generalized spherical cubature rule of arbitrary but odd degree is adopted to solve the above mentioned surface integral. The cubature rule adopted here was first proposed by Genz [50]. It was later utilized by Bin Jia and others [81] in the context of nonlinear filtering.

Theorem 2 *The spherical integral of the form* $I_{U_n}(\phi_{rZ}) = \int_{U_n} \phi(rZ) d\sigma(Z)$ *can be evaluated numerically for arbitrarily selected but odd degree as:*

$$I_{U_n,2m+1}(\phi_{rZ}) = \sum_{|p|} w_p \phi\{ru_p\}. \tag{4.19}$$

The above equation, $I_{U_n,(2m+1)}(\phi_{rZ})$ *represents spherical integration of the function* $\phi(rZ)$ *with the* $(2m+1)^{th}$ *degree spherical cubature rule, where* m *is any positive integer.* ru_p *and* w_p *are the cubature points and the corresponding weights whose expressions are given by,*

$$\{ru_p\} \triangleq \bigcup (\beta_1 ru_{p_1}, \beta_2 ru_{p_2}, \cdots, \beta_n ru_{p_n}) \tag{4.20}$$

and

$$w_p \triangleq 2^{-n_z(u_p)} \left(I_{U_n} \left(\prod_{i=1}^{n} \prod_{j=0}^{p_i-1} \frac{z_i^2 - u_j^2}{u_{p_i}^2 - u_j^2} \right) \right), \tag{4.21}$$

where p *is a set of non-negative numbers, defined as* $p = [p_1, p_2, \cdots, p_n]$, *and* $|p| = p_1 + p_2 + \cdots + p_n$; u_p *is also a set of non-negative numbers (not necessarily an integer);* $n_z(u_p)$ *gives the number of non-zero elements in* u_p; $\beta_i = \pm 1$ *and* $u_{p_i} = \sqrt{p_i/m}$.

Proof 2 *The proof of this theorem is provided in [50] and omitted here.*

∎

The intermediate weights, w_p, expressed in equation (4.21) can be evaluated using the following theorem.

Theorem 3 *The spherical integration of the form* $\int_{U_n} z_1^{\delta_1} z_2^{\delta_2} \cdots z_n^{\delta_n} d\sigma(Z)$ *where* $Z = [z_1, z_2, \cdots, z_n]^T$ *and* $\sigma(Z)$ *is the surface of unit hyper sphere,* U_n, *can be evaluated as [96]*

$$\int_{U_n} z_1^{\delta_1} z_2^{\delta_2} \cdots z_n^{\delta_n} dZ = 2 \frac{\Gamma((\delta_1 + 1)/2) \cdots \Gamma((\delta_n + 1)/2)}{\Gamma((|\delta| + n)/2)},$$

where $\Gamma(.)$ *represents the Gamma function and* $|\delta| = \delta_1 + \delta_2 + \cdots + \delta_n$.

Proof 3 *The proof of this theorem is provided in [81, 50] and omitted here.*

∎

4.7.3 High-degree cubature quadrature rule

Proposition 1 *Under the zero mean and unity covariance assumption, any integral in the form of (4.1) can be approximately evaluated as,*

$$I(\phi) = \frac{1}{\sqrt{(2\pi)^n}} \int_{r=0}^{\infty} \int_{U_n} [\phi(rZ)d\sigma(Z)]\, r^{n-1}e^{-r^2/2}dr$$

$$= \frac{1}{2\sqrt{\pi^n}} \sum_{i'=1}^{n'} \omega_{i'} \left[\sum_{|p|} w_p \phi\left\{ \sqrt{2\lambda_{i'}} u_p \right\} \right]. \tag{4.22}$$

Proof *Considering zero mean and unity covariance, the above integral becomes:*

$$I(\phi) = \frac{1}{\sqrt{(2\pi)^n}} \int_{r=0}^{\infty} \int_{U_n} [\phi(rZ)d\sigma(Z)]\, r^{n-1}e^{-r^2/2}dr. \tag{4.23}$$

Substituting equation (4.19) into equation (4.23), we get

$$I(\phi) = \frac{1}{\sqrt{(2\pi)^n}} \int_{r=0}^{\infty} \left[\sum_{|p|} w_p \phi\{ru_p\} \right] r^{n-1}e^{-r^2/2}dr.$$

Now to integrate the rest of the term, we use the Gauss-Laguerre quadrature formula described above. To cast the integration in the form of (4.6), let us substitute $t = r^2/2$. With this transformation the above integral becomes:

$$I(\phi) = \frac{1}{2\sqrt{\pi^n}} \int_{t=0}^{\infty} \left[\sum_{|p|} w_p \phi\left\{ \sqrt{2t} u_p \right\} \right] t^{n/2-1}e^{-t}dt. \tag{4.24}$$

Now, the above integral is a radial integral which is exactly in the form, as given in equation (4.6) for $\alpha = n/2 - 1$. Hence the Gauss-Laguerre quadrature rule can be applied for solving this integral. For i' number of quadrature points denoted as $\lambda_{i'}$ the integral (4.24) becomes:

$$I(\phi) = \frac{1}{2\sqrt{\pi^n}} \sum_{i'=1}^{n'} \omega_{i'} \left[\sum_{|p|} w_p \phi\left\{ \sqrt{2\lambda_{i'}} u_p \right\} \right].$$

∎

The mathematical formulation, described here, assumes zero mean and unity covariance. Like CQKF, this could easily be extended to any arbitrary mean and covariance.

It should be noted that the high-degree cubature quadrature rule constructed using the above theorems is fully symmetric. HDCQKF has two tuning parameters, namely m and n' which could be tuned by the designers to achieve the desired accuracy on an affordable computational load.

TABLE 4.2: Support points requirement of a n dimensional system for different degree of cubature and n' order quadrature rules.

Deg. of cub. rule	No. of support points
3	$2n \times n'$
5	$2n^2 \times n'$
7	$(4n^3 - 6n^2 + 8n) \times n'/3$
9	$(2n^4 - 8n^3 + 22n^2 - 10n) \times n'/3$
11	$(4n^5 - 30n^4 + 120n^3 + 30n^2 + 86n) \times n'/15$

4.7.4   Calculation of HDCQ points and weights

The steps for generating the points and weights for high-degree cubature quadrature Kalman filter are as follows:

- Determine all the possible sets of $p = [p_1, p_2, \cdots, p_n]$, such that $\mid p \mid = p_1 + p_2 + \cdots + p_n = m$, where p_i is the non-negative number.

- For every individual set of p, determine the set $u_p = [u_{p_1}, u_{p_2}, \cdots, u_{p_n}]$ where $u_{p_i} = \sqrt{p_i/m}$.

- Find out the set of the sample points for a given degree of cubature rule, $\xi = [\beta_1 u_{p_1}, \beta_2 u_{p_2}, \cdots, \beta_n u_{p_n}]$, where $\beta_i = \pm 1$.

- Calculate the intermediate weights, w_p, for corresponding cubature points using (4.21).

- Find the roots ($\lambda_{i'}$) of the n' order Chebyshev-Laguerre polynomial described in equation (4.7). Also calculate intermediate quadrature weights $\omega_{i'}$ using (4.8).

- For every set of p, determine the cubature quadrature points and corresponding weights as:

$$\xi_p = \bigcup (\beta_1 \sqrt{2\lambda_{i'}} u_{p_1}, \beta_2 \sqrt{2\lambda_{i'}} u_{p_2}, \cdots, \beta_n \sqrt{2\lambda_{i'}} u_{p_n}),$$

and $W_p = \dfrac{\omega_{i'} w_p}{2\sqrt{\pi^n}}$. The number of support points required for different degree of cubature rule is shown in Table 4.2.

4.7.5   Illustrations

Higher-order CQ points with a third degree spherical cubature rule

- For a 3rd degree spherical rule, $p_1 + p_2 + \cdots + p_n = 1$. Hence, $p = [1\ 0\ 0\ \cdots\ 0], [0\ 1\ 0\ \cdots\ 0], \cdots, [0\ 0\ 0\ \cdots\ 1]; u_p = [1\ 0\ 0\ \cdots\ 0], [0\ 1\ 0\ \cdots\ 0], \cdots, [0\ 0\ 0\ \cdots\ 1];$ and $\xi = [\pm 1\ 0\ 0\ \cdots\ 0], [0\ \pm 1\ 0\ \cdots\ 0], \cdots, [0\ 0\ 0\ \cdots\ \pm 1].$

- For a given specific p, the intermediate weight w_p will be given as

$$w_p = 2^{(-1)} \int_{U_n} z_1^2 d\sigma(Z)$$
$$= \frac{1}{2} \times 2 \frac{\Gamma((1+1)/2)\Gamma((0+1)/2) \cdots \Gamma((0+1)/2)}{\Gamma((n+2)/2)}$$
$$= \frac{\sqrt{\pi^n}}{n\Gamma(n/2)} = \frac{A_n}{2n},$$

where $A_n = 2\frac{\sqrt{\pi^n}}{\Gamma(n/2)}$ is the area of the unit hyper-sphere.

- Solve the n' order of the Chebyshev-Laguerre polynomial given in equation (4.7) to obtain $\lambda_{i'}$.

- Calculate quadrature weights $(\omega_{i'})$ corresponding to $\lambda_{i'}$ using (4.8).

- Finally the cubature quadrature points are $\xi_j = \sqrt{2\lambda_{i'}}\xi_i$.

- The corresponding weights associated with the cubature quadrature (CQ) points are:

$$W_j = \frac{\omega_{i'}w_p}{2\sqrt{\pi^n}} = \frac{1}{2n\Gamma(n/2)} \frac{n'!\Gamma(\alpha + n' + 1)}{\lambda_{i'}[\dot{L}_{n'}^{\alpha}(\lambda_{i'})]^2}, \qquad (4.25)$$

where $i = 1, 2, \cdots, 2n$, $i' = 1, 2, \cdots, n'$, $j = 1, 2, \cdots, 2nn'$.

Higher-order CQ points with the fifth degree spherical cubature rule

- For a 5th degree spherical rule, $p_1 + p_2 + \cdots + p_n = 2$. Hence $p = [2\ 0\ 0\ \cdots\ 0]$, $[0\ 2\ 0\ \cdots\ 0]$, $\cdots$, $[0\ 0\ 0\ \cdots\ 2]$, and $[1\ 1\ 0\ \cdots\ 0]$, $[1\ 0\ 1\ \cdots\ 0]$, $\cdots$, $[0\ 0\ \cdots\ 1\ 1]$.
  $u_p = [1\ 0\ 0\ \cdots\ 0]$, $[0\ 1\ 0\ \cdots\ 0]$, $\cdots$, $[0\ 0\ 0\ \cdots\ 1]$, and $[\sqrt{1/2}\ \sqrt{1/2}\ 0\ \cdots\ 0]$, $\cdots$ $[0\ 0\ \cdots\ \sqrt{1/2}\ \sqrt{1/2}]$.
  $\xi = [\pm 1\ 0\ 0\ \cdots\ 0]$, $[0\ \pm 1\ 0\ \cdots\ 0]$, $\cdots$ $[0\ 0\ 0\ \cdots\ \pm 1]$ and $[\pm\sqrt{1/2} \pm \sqrt{1/2}\ 0\ \cdots\ 0]$, $[\pm\sqrt{1/2}\ 0 \pm\sqrt{1/2}\ \cdots\ 0]$, $\cdots$, $[0\ 0\ \cdots \pm\sqrt{1/2} \pm\sqrt{1/2}]$.

- The intermediate weights,

$$w_{p=[2\ 0\ 0\ \cdots\ 0]} = w_{p=[0\ 2\ 0\ \cdots\ 0]} = \cdots = w_{p=[0\ 0\ 0\ \cdots\ 2]}$$
$$= 2^{(-1)} \int_{U_n} \frac{z_1^2 \left(z_1^2 - 1/2\right)}{1/2} d\sigma(z) = \frac{4-n}{2n(n+2)} A_n, \qquad (4.26)$$

and

$$w_{p=[1\ 1\ 0\ \cdots\ 0]} = w_{p=[1\ 0\ 1\ \cdots\ 0]} = \cdots = w_{p=[0\ 0\ \cdots\ 1\ 1]}$$
$$= 2^{(-2)} \int_{U_n} \frac{z_1^2 z_2^2}{\frac{1}{2}\frac{1}{2}} d\sigma(z) = \frac{A_n}{n(n+2)}. \qquad (4.27)$$

- Solve the n' order of the Chebyshev-Laguerre polynomial given in equation (4.7) to obtain $\lambda_{i'}$.

- Calculate the quadrature weights ($\omega_{i'}$), corresponding to $\lambda_{i'}$ using (4.8).

- Finally the CQ points are given as $\xi_j = \sqrt{2\lambda_{i'}}\xi_i$.

- The weights associated with CQ points are

$$W_j = \frac{\omega_{i'} w_p}{2\sqrt{\pi^n}} = \frac{n'!\Gamma(\alpha + n' + 1)}{\lambda_{i'}[\dot{L}_{n'}^{\alpha}(\lambda_{i'})]^2} \times \frac{w_p}{2\sqrt{\pi^n}}, \qquad (4.28)$$

where w_p is calculated with the expressions (4.26), and (4.27).

The points and their respective weights of a second order system for various odd degree cubature and up to the third order quadrature approximation are shown in Table 4.3. HDCQ points and their corresponding weights for a second order system with the 5th degree spherical cubature rule and 2nd order quadrature rule are shown in Figure 4.2. A MATLAB code to generate HDCQ points and weights is listed in Listing 4.2. It can generate points and weights for any order system with maximum 7th degree of cubature rule and any order quadrature rule.

```
1 % MATLAB program to generate HDCQ points and weights
2 % n: The dimension of the state
3 % n_dash: Order of Gauss-Laguerre quadrature
4 % m: Order of the cubature rule is (2m+1)
5 % This program can take m=1 or 2 or 3
6 % Credit: Kundan Kumar, IIT Patna
7 function [HDCQ_points, Weight] = HDCQ_points_wt(n,n_dash,m)
8 alpha = 0.5*n - 1;  %alpha
9 syms v; HDCQ_points=[]; Weight=[];
10 L(1,1)=1;mult=-n_dash*(n_dash+alpha);L(1,2)=mult;
11 for i=1:n_dash-1
12     mult=(-1)*mult*(n_dash-i)*(n_dash+alpha-i);
13     L(1,i+2)=mult/factorial(i+1);
14 end
15 o=L(1,:);  %Coefficient of chebyhev-Laguerre polynomial
16 r=roots(o); pv=(v^(alpha+n_dash))*exp(-v);
17 for i=1:n_dash
18     L1=diff(pv, sym('v')); pv=L1;
19 end
20 L2=((-1)^n_dash)*(v^(-alpha))*(exp(v))*pv;L3=diff(L2);
21 CL_poly_d=subs(L3,'v',r);  %Differentiated Chebyshev-Laguerre
       polynomial
22
23 switch(m)
24     case 1 %3rd degree spherical cubature rule ——————
25 [HDCQ_points,Weight]=cubquad_points(n,n_dash); % HDCQKF=CQKF
26
27     case 2 %5rd degree spherical cubature rule ——————
28 k=1;i=1;
29     while(k<n)
30         j=k+1;
```

```
31      while(j<n+1)
32          xi=zeros(n,1);xii=zeros(n,1);
33          xi(k,1)=1;  xi(j,1)=1;
34          xi1(:,i)=xi(:,1)/sqrt(2);
35          xii(k,1)=1;xii(j,1)=-1;
36          xi2(:,i)=xii(:,1)/sqrt(2);
37          i=i+1;j=j+1;
38      end
39      k=k+1;
40  end
41 s3=eye(n);
42  for j=1:n_dash
43      Temp1=sqrt(2*r(j,1));
44      HDCQ_points=[HDCQ_points Temp1*xi1 -Temp1*xi1 Temp1*xi2
            ...
45          -Temp1*xi2 Temp1*s3 -Temp1*s3];
46  end
47 Temp2=gamma(n/2)*n*(n+2);Temp3=gamma(alpha+n_dash+1);
48  for j=1:n_dash
49  Weight=[Weight (1/Temp2*(factorial(n_dash)*Temp3...
50      /(r(j,1)*CL_poly_d(j,1)^2)))*ones(1,2*n*(n-1)) ...
51      ((4-n)/(2*Temp2)*(factorial(n_dash)*...
52      Temp3/(r(j,1)*CL_poly_d(j,1)^2)))*ones(1,2*n)];
53  end
54
55      case 3 % For 7th degree spherical cubature————————
56 xi_1=eye(n);
57 b=[]; b=[b xi_1 -xi_1];
58 k=1; i=1;
59      while(k<n)
60          j=k+1;
61          while(j<n+1)
62              xi=zeros(n,1); xxi=zeros(n,1);xxxi=zeros(n,1);xxxxi=
                    zeros(n,1);
63              xi(k,1)=sqrt(2/3);xi(j,1)=sqrt(1/3);
64              xi_2(:,i)=xi(:,1);
65              xxi(k,1)=sqrt(2/3);xxi(j,1)=-sqrt(1/3);
66              xi_3(:,i)=xxi(:,1);
67              xxxi(k,1)=sqrt(1/3);xxxi(j,1)=sqrt(2/3);
68              xi_4(:,i)=xxxi(:,1);
69              xxxxi(k,1)=sqrt(1/3); xxxxi(j,1)=-sqrt(2/3);
70              xi_5(:,i)=xxxxi(:,1);
71              i=i+1;j=j+1;
72          end
73          k=k+1;
74      end
75 c=[];c=[c xi_2 -xi_2 xi_3 -xi_3 xi_4 -xi_4 xi_5 -xi_5];
76
77 if(n==2) % For a second order system————————————
78
79      for j=1:n_dash
80      HDCQ_points=[HDCQ_points sqrt(2*r(j,1))*b sqrt(2*r(j,1))*c];
81      end
82      for j=1:n_dash
83 Weight=[Weight (((2*n^2-15*n+43)* (factorial(n_dash)*gamma(alpha
        +n_dash+1)))...
```

```
84 /(4*gamma(n/2)*n*(n+2)*(n+4)*r(j,1)*CL_poly_d(j,1)^2))*ones
      (1,2*n)...
85 ((9*(5-n)*factorial(n_dash)*gamma(alpha+n_dash+1)))/(8*n*(n+2)
      *...
86 (n+4)*gamma(n/2)*r(j,1)*CL_poly_d(j,1)^2))*ones(1,4*n*(n-1))];
87    end
88
89 else    % For n greater than 2 ──────────────────────
90 k=1;i=1;
91    while(k<n)
92        j=k+1;
93        while(j<n+1)
94            l=j+1;
95            while(l<n+1)
96                xi=zeros(n,1);xyi=zeros(n,1);xxyi=zeros(n,1);xxyyi=
                    zeros(n,1);
97                xi(k,1)=1;xi(j,1)=1;xi(l,1)=1;
98                xi_6(:,i)=xi(:,1)/sqrt(3);
99
100               xyi(k,1)=1; xyi(j,1)=1; xyi(l,1)=-1;
101               xi_7(:,i)=xyi(:,1)/sqrt(3);
102
103               xxyi(k,1)=1; xxyi(j,1)=-1; xxyi(l,1)=1;
104               xi_8(:,i)=xxyi(:,1)/sqrt(3);
105
106               xxyyi(k,1)=1; xxyyi(j,1)=-1; xxyyi(l,1)=-1;
107               xi_9(:,i)=xxyyi(:,1)/sqrt(3);
108               i=i+1;l=l+1;
109           end
110           j=j+1;
111       end
112       k=k+1;
113   end
114   d=[];d=[d xi_6 -xi_6 xi_7 -xi_7 xi_8 -xi_8 xi_9 -xi_9];
115   for j=1:n_dash
116       HDCQ_points=[HDCQ_points sqrt(2*r(j,1))*b...
117           sqrt(2*r(j,1))*c sqrt(2*r(j,1))*d];
118   end
119   for j=1:n_dash
120       Weight=[Weight (((2*n^2-15*n+43)*(factorial(n_dash)...
121           *gamma(alpha+n_dash+1)))/(4*gamma(n/2)*n*(n+2)...
122           *(n+4)*r(j,1)*CL_poly_d(j,1)^2))*ones(1,2*n) ...
123           ((9*(5-n)*factorial(n_dash)*gamma(alpha+n_dash+1))...
124           /(8*n*(n+2)*(n+4)*gamma(n/2)*r(j,1)*CL_poly_d(j,1)^2))...
125       *ones(1,4*n*(n-1)) ((27*factorial(n_dash)*gamma(alpha+
              n_dash+1))...
126       /(8*gamma(n/2)*n*(n+2)*(n+4)*r(j,1)*CL_poly_d(j,1)^2))...
127           *ones(1,8*n*(n-1)*(n-2)/6)];
128   end
129 end
130 end
131% ────────────────────End of the code────────────────────
```

Listing 4.2: Generation of HDCQ points and weights.

TABLE 4.3: Points and weights for different degrees of cubature and higher order radial rules for a second order system

Degree of cubature $(2m+1)$	Radial order (n')	Points	Weights
3	1	(1.4142,0), (-1.4142,0), (0,1.4142), (0,-1.4142)	0.2500, 0.2500, 0.2500, 0.2500
	2	(2.6131,0), (-2.6131,0), (0,2.6131), (0,-2.6131), (1.0824,0), (-1.0824,0), (0,1.0824), (0,-1.0824)	0.0366, 0.0366, 0.0366, 0.0366, 0.2134, 0.2134, 0.2134, 0.2134
	3	(3.5468,0), (-3.5468,0), (0,3.5468), (0,-3.5468), (2.1421,0), (-2.1421,0), (0,2.1421), (0,-2.1421), (0.9119,0), (-0.9119,0), (0,0.9119), (0,-0.9119)	0.0026, 0.0026, 0.0026, 0.0026, 0.0696, 0.0696, 0.0696, 0.0696, 0.1778, 0.1778, 0.1778, 0.1778
5	1	(1.0000,1.0000), (-1.0000,-1.0000), (1.0000,-1.0000), (-1.0000,1.0000), (1.4142,0), (0,1.4142), (-1.4142,0), (0,1.4142)	0.1250, 0.1250, 0.1250, 0.1250, 0.1250, 0.1250, 0.1250, 0.1250
	2	(1.8478,1.8478), (-1.8478,-1.8478), (1.8478,-1.8478), (-1.8478,1.8478), (2.6131,0), (0,2.6131), (-2.6131,0), (0,-2.6131), (0.7654,0.7654), (-0.7654,0.7654), (0.7654,-0.7654), (-0.7654,-0.7654), (1.0824,0), (-1.0824,0), (0,1.0824), (0,1.0824)	0.0183, 0.0183, 0.0183, 0.0183, 0.0183, 0.0183, 0.0183, 0.0183, 0.1067, 0.1067, 0.1067, 0.1067, 0.1067, 0.1067, 0.1067, 0.1067
	3	(2.5080,2.5080), (-2.5080,2.5080), (2.5080,-2.5080), (-2.5080,-2.5080), (3.5468,0), (-3.5468,0), (0,3.5468), (0,-3.5468), (1.5147,1.5147), (-1.5147,1.5147), (1.5147,-1.5147), (-1.5147,-1.5147), (2.1421,0), (-2.1421,0), (0,2.1421), (0,-2.1421), (0.6448,0.6448), (-0.6448,0.6448), (0.6448,-0.6448), (-0.6448,-0.6448), (0.9119,0), (-0.9119,0), (0,0.9119), (0,-0.9119)	0.0013, 0.0013, 0.0013, 0.0013, 0.0013, 0.0013, 0.0013, 0.0013, 0.0348, 0.0348, 0.0348, 0.0348, 0.0348, 0.0348, 0.0348, 0.0348, 0.0889, 0.0889, 0.0889, 0.0889, 0.0889, 0.0889, 0.0889, 0.0889

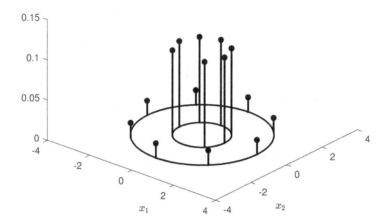

FIGURE 4.2: Plot of HDCQ points and weights for a 2D system with $m = n' = 2$.

4.7.6 High-degree cubature quadrature Kalman filter

In the previous subsection, we discussed how to generate high-degree cubature quadrature (HDCQ) points and corresponding weights. The points and the weights can be used to estimate the states of a nonlinear system. The methodology is known as high-degree cubature quadrature Kalman filter (HDCQKF). The algorithm of such a filter is the same as that described in Algorithm 8, apart from the fact that the deterministic support points and weights are replaced by HDCQ points. HDCQKF [142] is the most general form in the family of cubature filter.

Although any odd degree of cubature rule could be applied to solve the integral, designers restricted themselves to the fifth [185] and seventh degree of [186, 110] spherical cubature rule. A few interesting properties of the HDCQKF are listed below.

(i) For the 3rd degree spherical rule (i.e., $m = 1$) and the 1st order Gauss-Laguerre quadrature rule (i.e., $n' = 1$), the HDCQKF reduces to CKF [5].

(ii) For a special case of 3rd degree spherical (i.e., $m = 1$) and high-order Gauss-Laguerre quadrature rule (i.e., $n' > 1$), the HDCQKF merges with CQKF [17]. So we conclude, all the filters in the cubature family are a special and simplified case of the HDCQKF.

(iii) To increase the degree of accuracy of the HDCQKF, the degree of CQ approximation needs to be increased. The designer should also be aware that with the increase of degree of CQ approximation, the computational burden increases.

(iv) The main advantage of the HDCQKF is its computational accuracy compared to all existing filters in the cubature family (*viz.* CKF, CQKF, HD-CKF). Unfortunately, the accuracy comes at the cost of computational load. The computational burden of the HDCQKF is the highest compared to all other filters mentioned above. This may be considered as its demerit. However the computational burden of HDCQKF does not increase exponentially with the dimension of the system; hence HDCQKF is free from the curse of the dimensionality problem.

(v) The cubature filters are derivative free; that is, to implement them, neither a Jacobian nor Hessian matrix needs to be calculated. This may be considered as an added advantage from the computational point of view.

(vi) There are several avenues through which the work described in this chapter could be extended. The cubature filters could be reformulated for uncertain systems. In this respect risk sensitive [133] or H_∞ cost function [135] could be explored. The present method could also be extended to solve a nonlinear filtering problem with randomly delayed measurements [23].

(vii) For a predefined ordered cubature quadrature rule, the support points and their weights can be calculated and stored off line.

4.8 Simulation examples

In this section, the cubature filters are applied to solve two nonlinear filtering problems.

4.8.1 Problem 1

In this problem, we estimate the frequency and the amplitude of multiple superimposed sinusoids. For three sinusoids, the state vector becomes $\mathcal{X} = [f_1 \quad f_2 \quad f_3 \quad a_1 \quad a_2 \quad a_3]^T$, where f_i and a_i are the frequency in Hz and amplitude in Volts of the ith sinusoid. The state is considered to be following a random walk; hence the discrete state space equation will be

$$\mathcal{X}_k = I\mathcal{X}_{k-1} + \eta_k, \tag{4.29}$$

where I is a unitary matrix and η_k is process noise which is added to model the uncertainty that appeared in sinusoids. The process noise is normally distributed with zero mean and covariance Q, where, $Q = diag([\sigma_f^2 \ \sigma_f^2 \ \sigma_f^2 \ \sigma_a^2 \ \sigma_a^2 \ \sigma_a^2])$; $\sigma_f^2 = 0.051 \times 10^{-6}$ and $\sigma_a^2 = 0.8 \times 10^{-6}$.

The measurement equation is [31]

$$\mathcal{Y}_k = \left[\begin{array}{c} \sum_{j=1}^3 a_{j,k} \cos(2\pi f_{j,k} kT) \\ \sum_{j=1}^3 a_{j,k} \sin(2\pi f_{j,k} kT) \end{array} \right] + v_k,$$

where v_k is white Gaussian noise with zero mean and covariance $R = diag([0.09V^2 \ \ 0.09V^2])$ and T is the sampling time considered as 0.25 ms. In reality, the stream of measurement will come from the sensor attached to the system. Here, we have not used a hardware setup; rather we generate sensor measurements from the process dynamics and measurement equation. A typical program to generate sensor measurements is listed in Listing 4.3.

```
1 % Program to generate truth and measurement of Harmonic
     estimation problem
2 % The o/p of the program is an array of structure Ensm
3 % Ensm(m).x stores truth state for mth MC run
4 % Ensm(m).y stores measurement for mth MC run
5
6 n=6; %dimension of the systems
7 M=200; % Number of Monte Carlo run
8 Kmax=350; % Maximum step
9 t=1/5000; % Sampling time
10 F=eye(6); % Process matrix
11 sa=0.5*10^-6; sf=0.1*10^-6; sn=0.09; % noise parameters
12 Q=diag([sf sf sf sa sa sa]); % Process noise covariance
13 R=diag([0.09 0.09]); % Measurement noise covariance
14 for m=1:M  % Start of MC loop
15     Ensm(m).x(:,1)=[100;110;1000;5;4;3];
16     for k=1:Kmax % Start of time step loop
17         Ensm(m).x(:,k+1)=F*Ensm(m).x(:,k)+mvnrnd(zeros(6,1),Q)'; %
                True state
18 Ensm(m).y(:,k)=[Ensm(m).x(4,k)*cos(2*pi*Ensm(m).x(1,k)*k*t)+...
19 Ensm(m).x(5,k)*cos(2*pi*Ensm(m).x(2,k)*k*t)+Ensm(m).x(6,k)...
20     *cos(2*pi*Ensm(m).x(3,k)*k*t); Ensm(m).x(4,k)*sin(2*pi*...
21 Ensm(m).x(1,k)*k*t)+Ensm(m).x(5,k)*sin(2*pi*Ensm(m).x(2,k)*k*t)
        +...
22 Ensm(m).x(6,k)*sin(2*pi*Ensm(m).x(3,k)*k*t)]+mvnrnd(zeros(2,1),R
        )';
23     end
24 end
```

Listing 4.3: Generation of ensembled truth state and sensor measurements for problem 1.

The initial estimate is considered as normally distributed with mean $\mathcal{X}_0$ and covariance Σ_0. $\hat{\mathcal{X}}_0$ is the initial truth given as $[200 \ 1000 \ 2000 \ 5 \ 4 \ 3]^T$ and Σ_0 is the initial error covariance which is considered as $diag([200^2 \ 200^2 \ 200^2 \ 0.5 \ 0.5 \ 0.5])$ in standard unit. The states are estimated for 100 steps and the RMSE is calculated over 2000 Monte Carlo runs. The RMSE of frequency for the kth step is $(RMSE_k)^2 = (MSE_{1,k} + MSE_{2,k} + MSE_{3,k})/3$, where, for M number of Monte Carlo runs, $MSE_{i,k}$ is $MSE_{i,k} = \frac{1}{M} \sum_{j=1}^{M} (\mathcal{X}_{i,k,j} - \hat{\mathcal{X}}_{i,k,j})^2$. A similar calculation is done for the amplitude. A MATLAB program to estimate the states has been listed in Listing 4.4. The output of the program is an estimate of the state and error matrix for ensemble runs. From the error matrix RMSE can be calculated easily.

```
 1 % Function to estimate states of superimposed harmonics using
     CQKF
 2 % xes_in: Initialization of states of the estimator 6x1
 3 % P_in: Initial error covariance matrix 6x6
 4 % Meas: Array of m structure having two fields
 5 % Meas(m).x: All truth state for mth Ensemble run 6xkmax
 6 % Meas(m).y: All measurements for mth Ensemble run 2xkmax
 7 % m: number of Ensemble (MC run)
 8 % Meas is an array whose each element is 2 by kmax
 9 % noise: structure having two fields
10 % noise.Q=[sf*eye(3) zeros(3,3); zeros(3,3) sa*eye(3)];(sf, sa
11 % are scalar constant. A typical value: sa=0.5*10^-6;sf
     =0.1*10^-6;
12 % noise.R= Measurement noise cov, a typical value may be diag
     ([0.09 0.09]);
13 % Est= Array of m structure having two fields
14 % Est(m).x: Estimated value of state for mth Ensemble run 6xkmax
15 % Est(m).error: All estimation error for mth Ensemble run 6xkmax
16 function[Est]= harmonics(xes_in,P_in,Meas,noise)
17 n=length(xes_in); % Dimension of the system
18 M=length(Meas); % No of MC run
19 N=length(Meas(1).y); % Max time step i.e. kmax
20 n_dash=1; % n' for CQKF
21 [CQ_points,Weight]=cubquad_points(n,n_dash); % CQ points and wts
22 t=1/5000; % Sampling time
23 F=eye(6); P=P_in;x=xes_in;
24 %randn('seed',250) % Defining seed if necessary
25 for m=1:M
26
27    for k=1:N
28    S=chol(P,'lower'); %Cholesky factorization Mind the lower
29       for i=1:length(Weight) % Process upate
30       chi(:,i)=x+S*CQ_points(i,:)'; chinew(:,i)=F*chi(:,i);
31       end
32       meanpred=0;
33     for i=1:length(Weight)
34         meanpred=meanpred+Weight(i)*chinew(:,i);
35     end
36     Pnew=0;
37     for i=1:length(Weight) % Error cov calculation
38         Pnew=Pnew+Weight(i)*(chinew(:,i)-meanpred)*(chinew(:,i
            )-meanpred)';
39     end
40     Pnew1=Pnew+noise.Q;
41     S1=chol(Pnew1,'lower');
42
43     for i=1:length(Weight)
44     chii(:,i)=meanpred+S1*CQ_points(i,:)';
45     znew1(:,i)=[chii(4,i)*cos(2*pi*chii(1,i)*k*t)+chii(5,i)*cos
        (2*pi*...
```

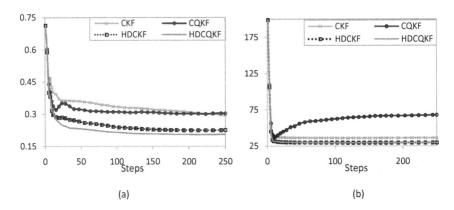

FIGURE 4.3: RMSE of (a) frequency in HZ (b) amplitude in volts [142].

```
46        chii(2,i)*k*t)+chii(6,i)*cos(2*pi*chii(3,i)*k*t);
47        chii(4,i)*sin(2*pi*chii(1,i)*k*t)+chii(5,i)*sin(2*pi*
            chii(2,i)...
48        *k*t)+chii(6,i)*sin(2*pi*chii(3,i)*k*t)];
49    end
50    zpred=0;
51      for i=1:length(Weight)
52          zpred=zpred+Weight(i)*znew1(:,i); % Expected
                measurement
53      end
54    Pnewm=0;
55      for i=1:length(Weight)
56        Pnewm=Pnewm+Weight(i)*(znew1(:,i)-zpred)*(znew1(:,i)-zpred
            )';
57      end
58    Pnewm1=Pnewm+noise.R; % Pyy
59    Pxznew=0;
60    for i=1:length(Weight)
61        Pxznew=Pxznew+Weight(i)*(chii(:,i)-meanpred)*(znew1(:,i)-
            zpred)';
62    end
63    K=Pxznew/Pnewm1; % Calculation of Kalman gain
64    x=meanpred+K*(Meas(m).y(:,k)-zpred);
65    Est(m).x(:,k)=x; Est(m).error(:,k)=x-Meas(m).x(:,k);
66    P=Pnew1-K*Pnewm1*K';
67 end
68
69 end
```

Listing 4.4: Estimation of state for problem 1.

From Figure 4.3, where the RMSE of frequency and amplitude have been plotted, we see that the RMSE obtained from HDCQKF is less compared to the CKF, CQKF, and HDCKF.

4.8.2 Problem 2

Let us consider a nonlinear system for which the process and measurement models are [27]:

$$\mathcal{X}_{k+1} = 20\cos(\mathcal{X}_k) + \eta_k, \tag{4.30}$$

and

$$\mathcal{Y}_k = \sqrt{1 + \mathcal{X}_k^T \mathcal{X}_k} + v_k. \tag{4.31}$$

Here, $\mathcal{X}_k$ is an n dimensional state vector at k^{th} instant of time. $\mathcal{Y}_k$ is the measurement at k^{th} step. η_k is the white Gaussian process noise. $\eta_k \sim \mathcal{N}(0_{n\times 1}, I_n)$, where $0_{n\times 1}$ denotes an n dimensional column matrix having all the elements equal to zero and I_n is an identity matrix of order n. $v_k \sim \mathcal{N}(0, 1)$ is white Gaussian measurement noise.

The initial truth states are considered as $\mathcal{X}_0 = 0.1 \times I_{n\times 1}$, while the filter is initialized with a value of $\hat{\mathcal{X}}_0$ and Σ_0, where $\hat{\mathcal{X}}_0 = 0_{n\times 1}$ and $\Sigma_0 = I_n$. The states are estimated using the CKF, CQKF, HDCKF, and HDCQKF. The accuracy of the estimators is compared in terms of root mean square error (RMSE). For M number of ensembles, the RMSE is defined as $RMSE_k = \sqrt{\frac{1}{M}\sum_{i=1}^{M} e_{i,k}^2}$, where the error $e_{i,k}$ is given as: $e_{i,k} = (\mathcal{X}_{i,k} - \hat{\mathcal{X}}_{i,k})$, $i = 1, 2, \cdots, M$, $k = 1, 2, \cdots$.

During simulation, a four dimensional system is considered and 500 Monte Carlo runs are used to evaluate the RMSE. The CQKF is implemented with the second order Gauss-Laguerre quadrature rule. The HDCKF is implemented with the fifth degree of cubature rule, and the HDCQKF is implemented with the fifth degree cubature and second order Gauss-Laguerre quadrature rule.

During the formulation of HDCQKF, we discussed that its accuracy is expected to be higher compared to all the other filters in the cubature family, *viz.* the CKF, the CQKF, and the HDCKF. Simulation results in Figure 4.4, which shows the RMSE plots for all the states and in Table 4.4 which discloses the averaged RMSE over time horizon, support our argument. From Figure 4.4, it could be seen that the RMSE obtained from HDCQKF is low for all the four states as compared to CKF, CQKF, and HDCKF. Table 4.4 also reveals the same fact.

The run times of CKF, CQKF, HDCKF and HDCQKF for the above problem in a personal computer with a 64-bit operating system, 4 GB RAM and 3.33 GHz i5 processor are 21.76, 30.63, 53.29 and 91.25 seconds respectively. So the computational burden for the HDCQKF is highest and almost four times that of CKF.

TABLE 4.4: RMSE averaged over time horizon (problem 2).

STATE	CKF	CQKF	HDCKF	HDCQKF
1	17.8527	17.6074	17.3249	16.1775
2	16.6974	16.2168	15.8685	14.6201
3	15.7352	15.0035	15.1503	14.2513
4	15.5288	14.7818	15.0433	14.1932

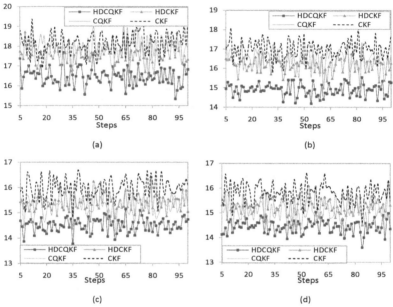

FIGURE 4.4: RMSE of (a) first state (b) second state (c) third state (d) fourth state [142].

4.9 Summary

In this chapter, the cubature quadrature Kalman filter and high-degree cubature quadrature Kalman filter are formulated. It has been shown that HDCQKF reduces to CQKF when the third degree spherical cubature rule is assumed. Moreover, CQKF reduces to CKF when the first order cubature rule is assumed.

Filters belonging to the cubature family have received warm acceptance from the designers. They are being used in different application areas such as target tracking [100, 78, 132], space applications [93], power systems [139], biomedical applications [65, 18] etc.

Further, researchers are engaged in designing various filters with the basic building block of the cubature rule of integration. Pakki and others combined an extended information filter and the cubature rule to develop a cubature information filter [119], and a high-degree cubature information filter [184]. A further square root version of it is also formulated [26]. Moreover, spherical simplex-radial versions of CKF [174], and CQKF [104] are available in recent literature. It is expected that in coming years the cubature filters will find more applications in various applied fields.

Chapter 5

Gauss-Hermite filter

5.1   Introduction 95
5.2   Gauss-Hermite rule of integration 96
     5.2.1   Single dimension 96
     5.2.2   Multidimensional integral 97
5.3   Sparse-grid Gauss-Hermite filter (SGHF) 99
     5.3.1   Smolyak's rule 100
5.4   Generation of points using moment matching method 104
5.5   Simulation examples 105
     5.5.1   Tracking an aircraft 105
5.6   Multiple sparse-grid Gauss-Hermite filter (MSGHF) 109
     5.6.1   State-space partitioning 109
     5.6.2   Bayesian filtering formulation for multiple approach ... 110
     5.6.3   Algorithm of MSGHF 111
     5.6.4   Simulation example 113
5.7   Summary 116

5.1 Introduction

In Chapters 3 and 4, chapter we have seen that the deterministic sample point filters assume that the prior and the posterior density functions are Gaussian (although they are not due to nonlinear process or measurement equation) and are characterized by the mean vector and the covariance matrix. Under such an assumption, during filtering, an integral appears in the form of $\int$ nonlinear function $\times$ Gaussian pdf which cannot be solved analytically for any arbitrary nonlinear function. In Chapter 4, this integral is approximately evaluated with cubature quadrature points. In this chapter, we shall use the Gauss quadrature method to evaluate the said integral. More specifically, as the weighting function here is exponential (being Gaussian pdf) the Gauss-Hermite quadrature rule will be useful. With this different method of numerical integration a new filtering algorithm was developed which is known as Gauss-Hermite filter (GHF). Although the Gauss-Hermite rule of integration is quite old in the literature [155], its application in filtering and estimation problems is recent and mainly due to the work of Ito and Xiong [74]. In this chapter, we shall discuss the Gauss-Hermite rule of integration and how it is solving the filtering problems.

Although GHF is quite an accurate nonlinear filter, it has certain computational disadvantages. The number of required support points increases exponentially with the dimension of the system. So, for a higher dimensional system, implementation of such a filter is computationally very expensive and sometimes impossible within the physical time limit for online applications. Here, we shall also discuss how such a problem can be partially solved.

5.2   Gauss-Hermite rule of integration

5.2.1   Single dimension

The Gauss-Hermite filter (GHF) makes use of the Gauss-Hermite Quadrature rule of integration [69] [96] [56]. Using the Gauss-Hermite quadrature rule, an integration of an arbitrary function over an exponential weighting function can be evaluated as a weighted sum over a set of sample points known as Gauss-Hermite quadrature points. From now on, we shall refer to them simply as quadrature points.

Consider an integral of single variable function $\phi(x)$,

$$I = \int_{-\infty}^{\infty} \phi(x) e^{-x^2} \, dx. \tag{5.1}$$

It can be evaluated numerically with a summation over N quadrature points

$$I \approx \sum_{j=1}^{N} \phi(q_j) w_j, \tag{5.2}$$

where q_j and w_j are the jth quadrature point and its corresponding weight respectively. The quadrature points and their weights can be evaluated using a symmetric tridiagonal matrix J with $J_{j,j+1} = \sqrt{j/2}$; $1 \leq j \leq (N-1)$. The quadrature points are obtained as $q_j = \sqrt{2}\Psi_j$, where Ψ_j is the jth eigenvalue of matrix J. The jth weight w_j can be defined as $w_j = k_{j1}^2$, where k_{j1} is the first element of the jth normalized eigenvector of J [6, 74, 25]. This method was actually proposed by Golub et al. [56] and was later used by Ito et al. [74], and Wu et al. [179] for filtering applications.

Illustration:

For a single dimensional integral, if we want to approximate with, say, 3 points, which we call a 3rd order Gauss-Hermite quadrature, our J matrix becomes, $J = \begin{bmatrix} 0 & \sqrt{1/2} & 0 \\ \sqrt{1/2} & 0 & 1 \\ 0 & 1 & 0 \end{bmatrix}$. Eigenvalues of the J matrix are $\Psi =$

$(-\sqrt{3/2}, 0, \sqrt{3/2})$. The quadrature points are $q_j = \sqrt{2}\Psi_j = (-\sqrt{3}, 0, \sqrt{3})$. The normalized eigenvectors are $k_1 = [-\sqrt{1/6} \ \sqrt{1/2} \ -\sqrt{1/3}]^T$; $k_2 = [-\sqrt{2/3} \ 0 \ -\sqrt{1/3}]^T$; $k_3 = [\sqrt{1/6} \ \sqrt{1/2} \ \sqrt{1/3}]^T$. The weights are $w = (1/6 \ 2/3 \ 1/6)$. So a 1D integral of the form (5.1) can be approximated with $\sum_i q_i w_i$, where the points and weights are given by $(q_1, w_1) = (-\sqrt{3}, 1/6)$, $(q_2, w_2) = (0, 2/3)$ and $(q_3, w_3) = (\sqrt{3}, 1/6)$. Similarly, points and weights can be calculated for any order Gauss-Hermite approximation.

5.2.2 Multidimensional integral

A multidimensional integral in the form of

$$I_N = \int_{R_n} f(s) \frac{1}{2\pi^{n/2}} e^{-\frac{1}{2}|s|^2} ds, \qquad (5.3)$$

can be approximately calculated with the help of the product rule applied on the above mentioned single dimensional quadrature points and weights. The above mentioned n dimensional integral can be approximately evaluated as

$$I_N = \sum_{j_1}^{N} \cdots \sum_{j_n}^{N} f(q_{j_1}, q_{j_2}, \cdots, q_{j_n}) w_{j_1} w_{j_2} \cdots w_{j_n}. \qquad (5.4)$$

If we take N number of single dimensional GH points, for a n dimensional system, a total of N^n number of quadrature points and their corresponding weights are required. As an example, a three point GH evaluation of an integral, for a three dimensional system requires twenty seven quadrature points and their weights. This can be expressed as q_i, q_j, q_k and $w_i w_j w_k$ respectively for $i = 1, 2, 3; j = 1, 2, 3; k = 1, 2, 3$ [141].

Illustration:

For a two dimensional system and three point GH quadrature rule, there will be $3^2 = 9$ number of support points and their weights. The support points will all be a combination of 1D Gauss-Hermite points and weights will be multiplication of corresponding weights. So the nine quadrature points and weights will be as follows (please note that the first bracket represents the points in 2D and in the second bracket weights are mentioned): $[(-\sqrt{3}, -\sqrt{3}), (1/6 \times 1/6)]$, $[(0, -\sqrt{3}), (2/3 \times 1/6)]$, $[(\sqrt{3}, -\sqrt{3}), (1/6 \times 1/6)]$, $[(-\sqrt{3}, 0), (1/6 \times 2/3)]$, $[(0, 0), (2/3 \times 2/3)]$, $[(\sqrt{3}, 0), (1/6 \times 2/3)]$, $[(-\sqrt{3}, \sqrt{3}), (1/6 \times 1/6)]$, $[(0, \sqrt{3}), (2/3 \times 1/6)]$, $[(\sqrt{3}, \sqrt{3}), (1/6 \times 1/6)]$. The above mentioned points and weights are represented in Figure 5.1. A code to obtain GH points and weights is provided in Listing 5.1.

```
1 % MATLAB program to generate GHKF points and weights
2 % n: The dimension of the states/system
3 % n1: No of points for univariate rule
4 % Code credit: Abhinoy Kumar Singh, IIT Indore
5 function [CP,W] = GH_points(n,n1)
```

```
6% ————Calculation of univariate Gauss quadrature————
7    J=zeros(n1,n1);
8 for i=1:n1-1
9    J(i,i+1)=sqrt(i/2); J(i+1,i)=sqrt(i/2);
10 end
11 [VJ,DJ]=eig(J); %Tridiagonal matrix
12 for i=1:n1 qp(:,i)=sqrt(2)*DJ(:,i); end
13 for i=1:n1
14    sum=0;
15    for j=1:n1 sum=sum+VJ(j,i)*VJ(j,i); end
16    W1(i)=VJ(1,i)*VJ(1,i)/sum; % Univariate weights
17 end
18% ————Generation of Multivariate points and weights ————
19 for k=1:n
20    q1=[];q2=[];
21    for i=1:n1
22       for j=1:n1^(n-k) q11(j)=qp(i,i); end
23       q1=[q1 q11]; q11=[];
24    end
25    for k1=1:n1^(k-1) q2=[q2 q1]; end
26 q3(k,:)=q2;
27 end
28 CP=q3; %Multivariate  points
29
30 for k=1:n
31    w1=[]; w2=[];
32    for i=1:n1
33       for j=1:n1^(n-k) w11(j)=W1(i); end
34       w1=[w1 w11]; w11=[];
```

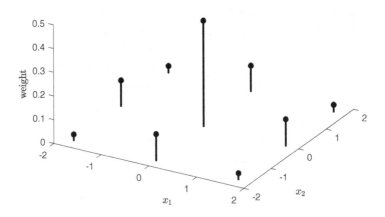

FIGURE 5.1: Plot of Gauss-Hermite points and weights for a 2D system with $N = 3$.

```
35      end
36        for k1=1:n1^(k-1) w2=[w2 w1]; end
37 w3(k,:)=w2;
38 end
39 W2=w3; W=[];
40 for i=1:n1^n
41     prod=1;
42     for j=1:n prod=prod*W2(j,i); end
43       W=[W prod]; % Weights associated
44 end
45 %————————————— End of the Program————————————
```

Listing 5.1: Generation of GH points and weights.

Here, we associate the 1D Gauss-Hermite quadrature points using the multi-plicative rule. In such association, the number of quadrature points require-ment increases exponentially with the increase in the dimension of systems. Hence the method suffers from the *curse of dimensionality problem*. To a certain extent, this problem can be minimized by neglecting the quadrature points on the diagonal because the weights associated with them are very small. So their contribution to the computation of integrals can be ignored. Further the readers should note the following points:

(i) The points and weight generating formula described above are for a Gaussian pdf with zero mean and unity covariance. For any arbitrary mean and covariance, the quadrature points could be calculated with a transformation with the mean and Cholesky factor of the covariance matrix.

(ii) The nonlinear filter which uses the Gauss-Hermite quadrature points and weights is known as the Gauss-Hermite filter (GHF). Once the points and weights are generated, the algorithm described in Algorithm 8 should be used to obtain the posterior and prior mean and covariance of states.

(iii) Similar to UKF and CKF, the Cholesky decomposition sometimes leads to negative definite covariance matrices, due to accumulated round off error associated with processing software. To get rid of this problem, square root GHF [4] may be used. The algorithm of the square root GHF is the same as the square root CKF as described in Algorithm 9.

5.3   Sparse-grid Gauss-Hermite filter (SGHF)

We have seen in the previous discussion that the GHF suffers from the curse of dimensionality problem. This means that the number of quadrature points increases exponentially with the dimension of the system. More specif-ically, the problem arises due to the multiplicative association between the

1D quadrature points. Instead of a multiplicative association, if we use the Smolyak's rule to generate multidimensional quadrature points, almost similar accuracy could be obtained from a smaller number of support points. When an estimator uses Smolyak's rule [53, 66] to generate quadrature points, the filtering algorithm is known as a Sparse-grid Gauss-Hermite filter (SGHF) [79, 80]. As the number of quadrature points is smaller with Smolyak's rule (compared to the multiplicative rule) the execution time of SGHF is accordingly less than that of the GHF.

The word sparse-grid filtering was first mentioned in [148]. Later, Heiss presented a detailed study of the univariate quadrature rule and its extension to multi-dimensional problems using sparse-grid theory. This gives thorough insight into the evolution of multivariate quadrature rules and the curse of dimensionality problem associated with it (product rule). The study shows that the integration rule using sparse-grid theory is exact for complete polynomials of a given order. Like the product rule, it combines the univariate quadrature rule, which makes it very general and easy to implement. The computational cost does not increase exponentially but at a considerably slower rate. This approach has now been widely used in numerical mathematics [21] [177].

5.3.1 Smolyak's rule

A numerical approximation for the following integral can be expressed as [177] [80]

$$
\begin{aligned}
I_{n,L}(f) &= \int_{\mathbb{R}^n} f(x)\mathcal{N}(x;0,I_n)dx \\
&\approx \sum_{q=L-n}^{L-1} (-1)^{L-1-q} C_{L-1-q}^{n-1} \sum_{\Xi \in \mathbf{N}_q^n} (I_{l_1} \otimes I_{l_2} \otimes \cdots \otimes I_{l_n})(f),
\end{aligned}
\tag{5.5}
$$

where the integral, $I_{n,L}$ represents the numerical evaluation of an n-dimensional system with the accuracy level L. This means that the approximation is exact for all the polynomials having degree up to $(2L-1)$. C stands for the binomial coefficient i.e., $C_k^n = n!/k!(n-k)!$. I_{l_j} is the single dimensional quadrature rule with the accuracy level $l_j \in \Xi$ and $\Xi \triangleq (l_1, l_2, \cdots, l_n)$, $\otimes$ stands for the tensor product and $\mathbf{N}_q^n$ is a set of possible values of l_j given as

$$
\begin{aligned}
\mathbf{N}_q^n &= \left\{ \Xi : \sum_{j=1}^{n} l_j = n+q \right\} && \text{for} \quad q \geq 0 \\
&= \varnothing && \text{for} \quad q < 0.
\end{aligned}
\tag{5.6}
$$

Equation (5.5) can be written as

$$I_{n,L}(f) \approx \sum_{q=L-n}^{L-1} (-1)^{L-1-q} C_{L-1-q}^{n-1} \sum_{\Xi \in \mathbf{N}_q^n} \sum_{q_{s_1} \in X_{l_1}} \sum_{q_{s_2} \in X_{l_2}}$$

$$\cdots \sum_{q_{s_n} \in X_{l_n}} f(q_{s_1}, q_{s_2}, \cdots, q_{s_n}) w_{s_1} w_{s_2} \cdots w_{s_n},$$

(5.7)

where X_{l_j} is the set of quadrature points for the single dimensional quadrature rule I_{l_j}, $[q_{s_1}, q_{s_2}, \cdots, q_{s_n}]^T$ is a sparse-grid quadrature (SGQ) point. $q_{s_j} \in X_{l_j}$ and w_{s_j} is the weight associated with q_{s_j}. Some SGQ points occur multiple times that could be counted once by adding their weight. The final set of the SGQ points is given as

$$X_{n,L} = \bigcup_{q=L-n}^{L-1} \bigcup_{\Xi \in \mathbf{N}_q^n} (X_{l_1} \otimes X_{l_2} \otimes \cdots \otimes X_{l_n}),$$

(5.8)

where $\bigcup$ represents union of the individual SGQ points.

Illustration:

For $n = 2$ and $L = 3$, values of q will be $q = 1$ and 2. Then from Eq. (5.6), for $q = 1$,

$$\mathbf{N}_q^n = \mathbf{N}_1^2 = l_1 + l_2 = 3 \Rightarrow (l_1, l_2) = \{(1, 2), (2, 1)\},$$

(5.9)

and for $q = 2$,

$$\mathbf{N}_q^n = \mathbf{N}_2^2 = l_1 + l_2 = 4 \Rightarrow (l_1, l_2) = \{(1, 3), (3, 1), (2, 2)\}.$$

(5.10)

Hence from Eq. (5.8), the point set can be defined as

$$X_{2,3} = (X_1 \otimes X_2) \cup (X_2 \otimes X_1) \cup (X_1 \otimes X_3) \cup (X_3 \otimes X_1) \cup (X_2 \otimes X_2).$$

(5.11)

Now, depending upon the univariate quadrature rule defined by the Gauss-Hermite quadrature (here $N = 3$), different point sets have entries as follows $X_1 = \{0\}$, $X_2 = \{-1.7, 0, 1.7\}$ and $X_3 = \{-2.8, -1.3, 0, 1.3, 2.8\}$. After determining the individual point sets, the total SGHF points for $n = 2$ can be defined as

$$X_{2,3} = \xi = \{(0, -1.7), (0, 0), (0, 1.7), (0, -2.8), (0, -1.3), (0, 1.3), (0, 2.8),$$
$$(-1.7, 0), (1.7, 0), (-1.7, -1.7), (-1.7, 1.7), (1.7, -1.7), (1.7, 1.7),$$
$$(-2.8, 0), (-1.3, 0), (1.3, 0), (2.8, 0)\}.$$

(5.12)

The points mentioned in the above illustration are shown in Figure 5.2. A MATLAB code to generate points and weights using Smolyak's rule is listed in Listing 5.2.

```
 1 % Program to generate sparse grid Gauss Hermite points and
        weights
 2 % n: Dimension of the states
 3 % n2: Accuracy level, L
 4 % points= Output points in n dimensional space;
 5 % wts=Weights of the points;
 6 % Code credit: Rahul Radhakrishnan of IIT Patna
 7
 8 function [points,wts] = GHsparse_points(n,n2)
 9 %Say n=2;  %n2=3;
10 % ———— Generation of 1D GH points and weights————
11 for k1=1:n2
12     n1=2*k1-1; J=zeros(n1,n1);
13     for i=1:n1-1          % Calculation of Tridiagonal matrix J
14         J(i,i+1)=sqrt(i/2);
15         J(i+1,i)=sqrt(i/2);
16     end
17 [VJ,DJ]=eig(J);
18     for i=1:n1 qp(k1,i)=sqrt(2)*DJ(i,i);end
```

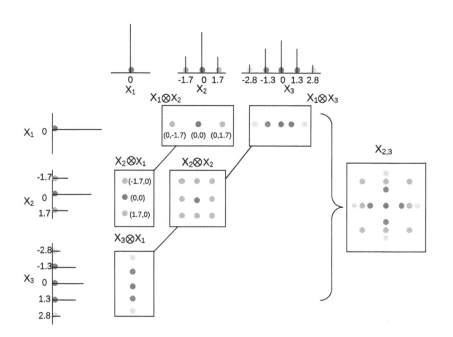

FIGURE 5.2: Sparse grid Gauss-Hermite points.

```
19 qp=round(qp*1e4)/1e4;
20    for i=1:n1
21        summ=0;
22        for j=1:n1 summ=summ+VJ(j,i)*VJ(j,i); end
23        W11(k1,i)=VJ(1,i)*VJ(1,i)/summ;   % normalized weight
24    end
25 end
26 % — Association in multidimension using Smolyak rule——————
27 CP=zeros(1,n);  W(1)=0;j1=2;
28 for q=n2-n:n2-1
29     x1=[]; d=n+q;  c = nchoosek(1:d+n-1,n-1);
30     m1 = size(c,1); t = ones(m1,d+n-1);
31     t(repmat((1:m1).',1,n-1)+(c-1)*m1) = 0;
32     u = [zeros(1,m1);t.'; zeros(1,m1)];
33     v = cumsum(u,1);
34     x = diff(reshape(v(u==0),n+1,m1),1).';
35     x(any(x==0,2),:)= [];
36     x1=[x1;x];
37     [row,col]=size(x1);
38     q4=[];w4=[];
39     for i1=1:row
40         q3=[];prod=1;
41         for i2=1:n prod=prod*(2*x1(i1,i2)-1);end
42         prod1=1;
43         for i=n:-1:1
44             q1=[]; q2=[];
45             for j=1:2*x1(i1,i)-1
46                 for k1=1:prod1
47                     q1=[q1 qp(x1(i1,i),j)];
48                 end
49             end
50             prod=prod/(2*x1(i1,i)-1);
51             for k2=1:prod q2=[q2 q1]; end
52             q3(i,:)=q2;prod1=prod1*(2*x1(i1,i)-1);
53         end
54         q4=[q4 q3];q3=q3';w3=[];prod=1;
55         for i2=1:n prod=prod*(2*x1(i1,i2)-1); end
56         prod1=1;
57         for i=n:-1:1
58             w1=[];w2=[];
59             for j=1:2*x1(i1,i)-1
60                 for k1=1:prod1 w1=[w1 W11(x1(i1,i),j)];end
61             end
62             prod=prod/(2*x1(i1,i)-1);
63             for k2=1:prod w2=[w2 w1]; end
64             w3(i,:)=w2; prod1=prod1*(2*x1(i1,i)-1);
65         end
66         w4=[w4 w3];  w3=w3';[mm1,nn1]=size(q3);
67         for i=1:mm1
68             [mm2,nn2]=size(CP);
69             while mm2>0
70                 tf=ismember(q3(i,:),CP(mm2,:),'rows');
71                 if(tf==1) j2=mm2; mm2=-1; end
72                 mm2=mm2-1;
73             end
74             if(tf==0)
75                 prod3=1;
```

```
76      for j=1:nn1 prod3=prod3*w3(i,j);end
77      CP=[CP;q3(i,:)];
78      W(j1)=((-1)^(n2-1-q))*(factorial(n-1)...
79          /(factorial(n2-1-q)*factorial(n-n2+q)))*
            prod3;
80      j1=j1+1;
81  else
82      prod4=1;
83      for j=1:nn1 prod4=prod4*w3(i,j); end
84      W(j2)=W(j2)+((-1)^(n2-1-q))*(factorial(n-1)...
85          /(factorial(n2-1-q)*factorial(n-n2+q)))*
            prod4;
86      end
87  end
88  end
89
90 end
91 points=CP'; wts=W;
92% ───────────────── End of the code ─────────────────
```

Listing 5.2: Generation of sparse Gauss-Hermite quadrature points and weights.

It is clear from the above formulation and illustration that the total number of sparse-grid quadrature points does not increase exponentially with the dimension. Hence, the *curse of dimensionality problem* of GHF is alleviated in this formulation. A relation between the sparse-grid quadrature filter and cubature Kalman filter was studied in [82]. It was found that the projection of the sparse-grid quadrature rule results in arbitrary degree cubature rules. Application of SGHF to various real life problems was made in [79] [77] etc.

Other than Gauss-Hermite points and the tensor operation between them, there exist a few other methods for generating quadrature points and weights. Among them Clenshaw-Curtis [168], Gauss-Kronrod Quadrature [99, 22], extended Gauss (Patterson) quadrature [53, 51], adaptive Gauss quadrature rule [54] are important. Hermann Singer introduces generalization of the Gauss-Hermite filter in [140] which is claimed to be better than GHF. Very recently, the adaptive Gauss quadrature rule is used to develop a new nonlinear filtering method [145] which is computationally faster and more accurate than the sparse grid Gauss-Hermite filter.

5.4   Generation of points using moment matching method

Univariate quadrature points and weights can alternatively be determined using a moment matching method. SGHF and GHF can also use this method for determining quadrature points [80] [6]. The general moment matching formula used for a one dimensional Gaussian type integral is

$$M_j = \int_{-\infty}^{\infty} x^j \mathcal{N}(x; 0, 1) dx = \sum_{i=1}^{N_u} q_i^j w_i, \qquad (5.13)$$

where M_j is the jth order moment and N_u represents the number of univariate quadrature points. q_i and w_i give the univariate quadrature points and weights respectively. The quadrature rule should be exact for all univariate polynomials of order up to $N_u - 1$. So, q_i and w_i should satisfy the equation:

$$\begin{pmatrix} 1 & 1 & \cdots & 1 \\ q_1 & q_2 & \cdots & q_m \\ \vdots & \vdots & \ddots & \vdots \\ q_1^{m-1} & q_2^{m-1} & \cdots & q_m^{m-1} \end{pmatrix} \begin{pmatrix} w_1 \\ w_2 \\ \vdots \\ w_m \end{pmatrix} = \begin{pmatrix} M_0 \\ M_1 \\ \vdots \\ M_{m-1} \end{pmatrix} \qquad (5.14)$$

If the number of quadrature points is m, we have $2m$ number of unknowns including m number of quadrature points and their corresponding m weights. But there are only m number of moment equations available. So, some authors select the quadrature points arbitrarily and calculate the corresponding weights by using moment equations [80], while some authors choose the quadrature points as the zeros of the Hermite polynomial [121]. However this method may suffer from numerical instability [6]. A better method for selecting the quadrature points and their corresponding weights was demonstrated by Golub et al. in [56] and later utilized by Arasaratnam et al. in [6].

5.5   Simulation examples

5.5.1   Tracking an aircraft

The maneuver of a civilian aircraft generally follows a constant turn rate. Knowledge about the speed and the turn rate during maneuvers is extremely important for air traffic control. In this subsection, a problem of maneuvering target tracking with a constant but unknown turn rate has been formulated. However, to some extent the model could also be used for varying turn rate as the noise is incorporated to capture the variability. The target, assumed to be maneuvering with a constant turn rate, is popularly known as coordinated turn in avionics vocabulary [13]. The coordinated turn model, adopted for target motion is summarized in [102] and is well described in [13]. In recent years, Arasaratnam et al. [5] and Bin Jia et al. [81] have adopted this problem to compare the accuracy of their proposed algorithms with existing methods.

To formulate the problem, we assume an object is maneuvering with a constant turn rate in a plane parallel to the ground i.e., during maneuvers the height of the vehicle remains constant. If the turn rate is a known constant, the process model remains linear. However, a constant but unknown turn rate, which needs to be estimated, forces the process model to a set of nonlinear equations. The equation of motion of an object in a plane (x, y) following a coordinated turn model could be described as

$$\ddot{x} = -\Omega \dot{y} \tag{5.15}$$

$$\ddot{y} = \Omega \dot{x} \tag{5.16}$$

$$\dot{\Omega} = 0, \tag{5.17}$$

where x, y represent the position in x and y direction respectively. Ω is the angular rate which is a constant. In navigation convention, $\Omega < 0$ implies a counter clockwise turn. State space representation of the above equations is

$$\dot{\mathcal{X}} = A\mathcal{X} + w, \tag{5.18}$$

where $\mathcal{X}$ is a state vector defined as $\mathcal{X} = [x \ \dot{x} \ y \ \dot{y} \ \Omega]^T$. The process noise is added to incorporate the uncertainties in the process equation, arising due to wind speed, variation in turn rate, change in velocity etc.

The target dynamics is discretized to obtain the discrete process equation as

$$\mathcal{X}_{k+1} = \phi_k \mathcal{X}_k + w_k, \tag{5.19}$$

where

$$
\phi_k = \begin{bmatrix}
1 & \dfrac{\sin(\Omega_{k-1}T)}{\Omega_{k-1}} & 0 & -\dfrac{1 - \cos(\Omega_{k-1}T)}{\Omega_{k-1}} & 0 \\
0 & \cos(\Omega_{k-1}T) & 0 & -\sin(\Omega_{k-1}T) & 0 \\
0 & \dfrac{1 - \cos(\Omega_{k-1}T)}{\Omega_{k-1}} & 1 & \dfrac{\sin(\Omega_{k-1}T)}{\Omega_{k-1}} & 0 \\
0 & \sin(\Omega_{k-1}T) & 0 & \cos(\Omega_{k-1}T) & 0 \\
0 & 0 & 0 & 0 & 1
\end{bmatrix}.
$$

In general, the nonlinear measurement equation could be written as

$$\mathcal{Y}_k = \gamma(\mathcal{X}_k). \tag{5.20}$$

In this problem, we assume the range and the bearing angle both are available from the measurement. So the nonlinear function $\gamma(.)$ becomes

$$\gamma(\mathcal{X}_k) = \begin{bmatrix} \sqrt{x_k^2 + y_k^2} \\ \mathrm{atan2}(y_k, x_k) \end{bmatrix} + v_k, \tag{5.21}$$

where *atan2* is the four quadrant inverse tangent function. Both w_k and v_k are white Gaussian noise of zero mean and Q and R covariance respectively and T is sampling time.

The maneuvering target tracking problem formulated above has been solved in the MATLAB environment by using the UKF, CKF, GHF and

SGHF. Experimentation has been done by considering $\kappa = -2$ for the UKF, 3-points GHF and 3rd-degree of accuracy for the SGHF. We consider the process noise Q as

$$
Q = \begin{bmatrix} g\dfrac{T^3}{3} & g\dfrac{T^2}{2} & 0 & 0 & 0 \\ g\dfrac{T^2}{2} & gT & 0 & 0 & 0 \\ 0 & 0 & g\dfrac{T^3}{3} & g\dfrac{T^2}{2} & 0 \\ 0 & 0 & g\dfrac{T^2}{2} & gT & 0 \\ 0 & 0 & 0 & 0 & 0.009T \end{bmatrix}, \tag{5.22}
$$

where the sampling time T is taken as 0.5 second and g is some constant given as $g = 0.1$. R is considered as $\mathrm{diag}([\sigma_r^2 \quad \sigma_t^2])$ where $\sigma_r = 120\mathrm{m}$ and $\sigma_t = \sqrt{70}\mathrm{mrad}$.

The initial truth value is considered as

$$
\mathcal{X}_0 = [1000\mathrm{m} \ 30\mathrm{m/s} \ 1000\mathrm{m} \ 0\mathrm{m/s} \ -3^\circ/s]^T,
$$

while the initial estimate of the covariance is

$$
\Sigma_0 = \mathrm{diag}([200\mathrm{m}^2 \ 20\mathrm{m}^2/\mathrm{s}^2 200\mathrm{m}^2 \ 20\mathrm{m}^2/\mathrm{s}^2 \ 100\mathrm{mrad}^2/\mathrm{s}^2]).
$$

The initial estimate is considered to be a normally distributed random number with mean $\mathcal{X}_0$ and covariance Σ_0. The simulation is performed for 50 seconds and the result is compared by evaluating the RMSE of position, velocity and turn rate for 50 independent Monte Carlo runs.

The RMSE plots of range, resultant velocity and turn rate are shown in Figure 5.3 to Figure 5.5. The RMSE plots show that the accuracy of the GHF

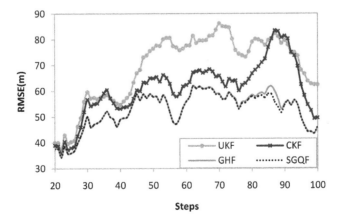

FIGURE 5.3: RMSE of range out of 50 MC runs.

and SGHF are better than the UKF and CKF. The computational time for the CKF, GHF and SGHF are noticed as 1.32, 11.45 and 10.16 times higher than the UKF.

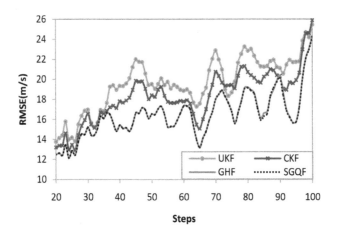

FIGURE 5.4: RMSE of resultant velocity out of 50 MC runs.

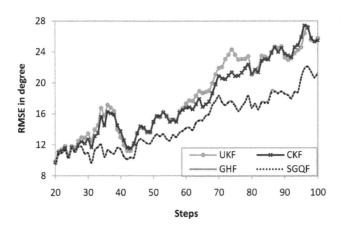

FIGURE 5.5: RMSE of turn rate out of 50 MC runs.

5.6 Multiple sparse-grid Gauss-Hermite filter (MSGHF)

In this approach [127], a combination of SGHF and multiple state-space partitioning technique [31] is implemented, so that the resultant filter's performance is comparable with other variants of GHFs at a lower computational cost. SGHF itself is efficient in reducing the computational cost since the number of points required increases only as a polynomial function of the dimension of the system. When applied with a state-space partitioning technique, it further reduces the computational cost, without hampering the accuracy measures considerably.

5.6.1 State-space partitioning

The process equation mentioned in (1.19) can be partitioned into S different subspaces, having the same or different dimensions. The partitioned process equation can be expressed as

$$
\begin{pmatrix} \mathcal{X}_k^{(1)} \\ \mathcal{X}_k^{(2)} \\ \vdots \\ \mathcal{X}_k^{(S)} \end{pmatrix} = \begin{pmatrix} \phi_{k-1}^{(1)}(\mathcal{X}_{k-1}^{(1)}, \mathcal{X}_{k-1}^{(-1)}) \\ \phi_{k-1}^{(2)}(\mathcal{X}_{k-1}^{(2)}, \mathcal{X}_{k-1}^{(-2)}) \\ \vdots \\ \phi_{k-1}^{(S)}(\mathcal{X}_{k-1}^{(S)}, \mathcal{X}_{k-1}^{(-S)}) \end{pmatrix} + \begin{pmatrix} \eta_{k-1}^{(1)} \\ \eta_{k-1}^{(2)} \\ \vdots \\ \eta_{k-1}^{(S)} \end{pmatrix}. \tag{5.23}
$$

Here, $\phi_{k-1}^{(s)}(\mathcal{X}_{k-1}) : R^{n_\mathcal{X}} \to R^{n_\mathcal{X}^{(s)}}$ represents the function corresponding to each partitioned subspace, with $s = 1, \cdots, S$, where S is the total number of subspaces and $\Sigma_{s=1}^{S} n_\mathcal{X}^{(s)} = n_\mathcal{X}$. Each function is represented as $\phi_{k-1}^{(s)}(\mathcal{X}_{k-1}^{(s)}, \mathcal{X}_{k-1}^{(-s)})$ instead of $\phi_{k-1}^{(s)}(\mathcal{X}_{k-1}^{(s)})$ to highlight the clear-cut nature of state-space partitioning. Along with the state-space, the s-th process noise is also described as $\eta_{k-1}^{(s)} \sim \mathcal{N}(0, Q_{k-1}^{(s)})$, where $Q_{k-1}^{(s)}$ can be constructed from Q_{k-1} with the entries corresponding to the s-th subspace. The partitioned process noise $\eta_{k-1}^{(s)}$ can be equivalent to original process noise η_{k-1}, if there are no correlated terms. In this approach, the process noise of different subspaces is taken into consideration by its corresponding subspace only. This neglects the possible cross-correlations among subspace process noise vectors and hence stands as a source of error.

Thus, the s-th subspace process equation can be obtained as

$$
\mathcal{X}_k^{(s)} = \phi_{k-1}^{(s)}(\mathcal{X}_{k-1}^{(s)}) + \eta_{k-1}^{(s)},
$$

where $\phi_{k-1}^{(s)}(\mathcal{X}_{k-1}^{(s)}) : R^{n_\mathcal{X}^{(s)}} \to R^{n_\mathcal{X}^{(s)}}$. The state-space partitioning technique becomes much simplified when we consider independent subspaces i.e., when there is no interconnection among subspaces (*e.g:* a random walk problem) and when there is no correlation among different subspace process noises.

5.6.2 Bayesian filtering formulation for multiple approach

The filters which take care of their corresponding subspaces are mainly used for estimating the marginal posterior probability densities, $p(x_k^{(s)} | x_k^{(-s)}, y_{1:k})$. Hence, the Bayesian recursive solution mentioned in (1.29) and (1.31) can be modified as [31]

$$p(\mathcal{X}_k^{(s)} | \mathcal{X}_{k-1}^{(-s)}, \mathcal{Y}_{1:k-1}) = \int p(\mathcal{X}_k^{(s)} | \mathcal{X}_{k-1}^{(-s)}, \mathcal{X}_{k-1}^{(s)}) p(\mathcal{X}_{k-1}^{(s)} | \mathcal{X}_{k-1}^{(-s)}, \mathcal{Y}_{1:k-1}) d\mathcal{X}_{k-1}^{(s)},$$
(5.24)

and

$$p(\mathcal{X}_k^{s} | \mathcal{X}_k^{(-s)}, \mathcal{Y}_{1:k}) \propto p(\mathcal{Y}_k | \mathcal{X}_k^{(s)}, \mathcal{X}_k^{(-s)}) p(\mathcal{X}_k^{(s)} | \mathcal{X}_{k-1}^{(-s)}, \mathcal{Y}_{1:k-1}).$$
(5.25)

Considering that all the densities follow normal distribution, the prior and posterior probability densities can be written as

$$p(\mathcal{X}_k^{(s)} | \mathcal{X}_{k-1}^{(-s)}, \mathcal{Y}_{1:k-1}) = \mathcal{N}(\hat{\mathcal{X}}_{k|k-1}^{(s)}, \Sigma_{k|k-1}^{(s)})$$
(5.26)

and

$$p(\mathcal{X}_k^{(s)} | \mathcal{X}_k^{(-s)}, \mathcal{Y}_{1:k}) = \mathcal{N}(\hat{\mathcal{X}}_{k|k}^{(s)}, \Sigma_{k|k}^{(s)}).$$
(5.27)

The problem of estimating the posterior mean and covariance, $(\hat{\mathcal{X}}_{k|k}^{(s)}, \Sigma_{k|k}^{(s)})$, at any instant k, from the noisy measurements and from the prior mean and covariance $(\hat{\mathcal{X}}_{k|k-1}^{(s)}, \Sigma_{k|k-1}^{(s)})$, can be addressed by the following set of equations.

Prediction:

$$\hat{\mathcal{X}}_{k|k-1}^{(s)} = \mathbb{E}[\mathcal{X}_k^{(s)} | \mathcal{X}_{k-1}^{(-s)}, \mathcal{Y}_{1:k-1}]$$
$$= \int \phi(\mathcal{X}_{k-1}^{(s)}, \mathcal{X}_{k-1}^{(-s)}) p(\mathcal{X}_{k-1}^{(s)} | \mathcal{X}_{k-1}^{(-s)}, \mathcal{Y}_{1:k-1}) d\mathcal{X}_{k-1}^{(s)}$$
(5.28)

$$\Sigma_{\mathcal{X},k|k-1}^{(s)} = \mathbb{E}[(\mathcal{X}_k^{(s)} - \hat{\mathcal{X}}_{k|k-1}^{(s)})(\mathcal{X}_k^{(s)} - \hat{\mathcal{X}}_{k|k-1}^{(s)})^T | \mathcal{X}_{k-1}^{(-s)}, \mathcal{Y}_{1:k-1}]$$
$$= \int \phi^2(\mathcal{X}_{k-1}^{(s)}, \mathcal{X}_{k-1}^{(-s)}) p(\mathcal{X}_{k-1}^{(s)} | \mathcal{X}_{k-1}^{(-s)}, \mathcal{Y}_{1:k-1}) d\mathcal{X}_{k-1}^{(s)}$$
$$- (\hat{\mathcal{X}}_{k|k-1}^{(s)})^2 + Q_{k-1}^{(s)}$$
(5.29)

Update:

$$\hat{\mathcal{X}}_{k|k}^{(s)} = \hat{\mathcal{X}}_{k|k-1}^{(s)} + K_k^{(s)}(\mathcal{Y}_k - \hat{\mathcal{Y}}_{k|k-1}^{(s)})$$
(5.30)

$$\Sigma_{k|k}^{(s)} = \Sigma_{k|k-1}^{(s)} - K_k^{(s)} \Sigma_{yy,k|k-1}^{(s)} (K_k^{(s)})^T$$
(5.31)

The measurement prediction, innovation covariance, cross-covariance and Kalman gain can be obtained as:

$$\hat{\mathcal{Y}}_{k|k-1}^{(s)} = \int \gamma(\mathcal{X}_k^{(s)}, \mathcal{X}_k^{(-s)}) p(\mathcal{X}_k^{(s)} | \mathcal{X}_k^{(-s)}, \mathcal{Y}_{1:k-1}) d\mathcal{X}_k^{(s)},$$
(5.32)

$$\Sigma^{(s)}_{\mathcal{YY},k|k-1} = \int \gamma^2(\mathcal{X}^{(s)}_k, \mathcal{X}^{(-s)}_k)p(\mathcal{X}^{(s)}_k | \mathcal{X}^{(-s)}_k, \mathcal{Y}_{1:k-1})d\mathcal{X}^{(s)}_k$$
$$- (\hat{\mathcal{Y}}^{(s)}_{k|k-1})(\hat{\mathcal{Y}}^{(s)}_{k|k-1})^T + R_k, \tag{5.33}$$

$$\Sigma^{(s)}_{\mathcal{XY},k|k-1} = \int \mathcal{X}^{(s)}_k \gamma^T(\mathcal{X}^{(s)}_k, \mathcal{X}^{(-s)}_k)p(\mathcal{X}^{(s)}_k | \mathcal{X}^{(-s)}_k, \mathcal{Y}_{1:k-1})d\mathcal{X}^{(s)}_k$$
$$- \hat{\mathcal{X}}^{(s)}_{k|k-1}(\hat{\mathcal{Y}}^{(s)}_{k|k-1})^T, \tag{5.34}$$

and

$$K^{(s)}_k = \Sigma^{(s)}_{\mathcal{XY},k|k-1}(\Sigma^{(s)}_{\mathcal{YY},k|k-1})^{-1}. \tag{5.35}$$

Pdfs mentioned in the above integrals are assumed to be Gaussian and the integrals could be evaluated with the sparse grid support points and weights. Here, some important points of MSGHF are summarized below.

(i) The partitioning technique can only be applied to problems where the function $\phi(\cdot)$ is only diagonally occupied.

(ii) During partition, cross correlation terms of the process covariance matrix may get neglected. Avoiding this information makes the estimation process prone to errors.

(iii) As mentioned in (5.29), $\Sigma^{(s)}_{k|k-1}$, the prior error covariance of each subspace is determined using $Q^{(s)}_{k-1}$, the corresponding subspace process noise covariance. Since $Q^{(s)}_{k-1}$ carries only the noise affecting each subspace instead of full process noise covariance, these errors are propagated through the filtering algorithm.

(iv) Since S subspaces are dealt with separately using dedicated filters, it doesn't mean that we are dealing with S independent estimation problems. This is clear from the fact that the observations still may contain the full state space which comes into the picture when innovations are used for S-th filter prediction.

(v) State partitioning, which is set by the practitioner, should depend on the amount of nonlinearity involved, available computational resources and also on the application. Given a state-space for partitioning, the preferred way is to group the states which are highly correlated into a particular subspace and putting the uncorrelated ones into different subspaces.

5.6.3   Algorithm of MSGHF

The algorithm of the MSGHF is as follows.

Step 1 Generation of sparse-grid Gauss-Hermite quadrature points and weights

- Specify the state-space partitioning. Generate univariate quadrature point sets X_{l_j} and weights, w_j for the desired accuracy level using Golub's technique [56].

- Extend them to multidimensional space using Smolyak's rule and obtain the SGH points q_{s_j} and weights w_{s_j}, where $j = 1, 2, \cdots, P_s$, P_s is the number of point in subspace s.

Step 2 Filter Initialization

- Initialize the independent filters with their corresponding initial estimate $\hat{\mathcal{X}}_{0|0}^{(s)}$ and initial covariance $\Sigma_{\mathcal{X},0|0}^{(s)}$ and noise covariances $\hat{Q}_k^{(s)}$ and $\hat{R}_k$.

Step 3 Prediction step or propagation of SGH points

- Perform Cholesky decomposition of posterior error covariance to find its square-root, such that

$$\Sigma_{\mathcal{X},k-1|k-1}^{(s)} = \zeta_{\mathcal{X},k-1|k-1}^{(s)} \zeta_{\mathcal{X},k-1|k-1}^{(s)T}.$$

- Evaluate SGH points with a given mean and covariance:

$$\mathcal{X}_{j,k-1|k-1}^{(s)} = \zeta_{\mathcal{X},k-1|k-1}^{(s)} \xi_j^{(s)} + \hat{\mathcal{X}}_{k-1|k-1}^{(s)}$$

- Propagate each one through the corresponding state transition function $\phi(\cdot)$:

$$\mathcal{X}_{j,k|k-1}^{(s)} = \phi(\mathcal{X}_{j,k-1|k-1}^{(s)})$$

- The subspace mean can be predicted by computing the weighted mean of the propagated sigma-points:

$$\hat{\mathcal{X}}_{k|k-1}^{(s)} = \sum_{j=1}^{Ps} w_j^{(s)} \mathcal{X}_{j,k|k-1}^{(s)}$$

- Prior covariance can be computed as

$$\Sigma_{\mathcal{X},k|k-1}^{(s)} = \sum_{j=1}^{P_s} w_j^{(s)} (\mathcal{X}_{j,k|k-1}^{(s)} - \hat{\mathcal{X}}_{k|k-1}^{(s)})(\mathcal{X}_{j,k|k-1}^{(s)} - \hat{\mathcal{X}}_{k|k-1}^{(s)})^T + \hat{Q}_k^{(s)}.$$

Step 4 Measurement update

- Perform the Cholesky decomposition of prior covariance:

$$\Sigma_{\mathcal{X},k|k-1}^{(s)} = \zeta_{\mathcal{X},k|k-1}^{(s)} \zeta_{\mathcal{X},k|k-1}^{(s)T}$$

- Evaluate the SGH points with a given mean and covariance:

$$\tilde{\mathcal{X}}_{j,k|k-1}^{(s)} = \zeta_{\mathcal{X},k|k-1}^{(s)}\xi_j^{(s)} + \hat{\mathcal{X}}_{k|k-1}^{(s)}$$

- Propagate the SGH points through the given measurement function:

$$Y_{j,k|k-1}^s = \gamma(\tilde{\mathcal{X}}_{j,k|k-1}^{(s)}, \hat{\mathcal{X}}_{j,k|k-1}^{(s)})$$

- The predicted measurement can be obtained as

$$\hat{Y}_{k|k-1}^{(s)} = \sum_{j=1}^{P_s} w_j^{(s)} Y_{j,k|k-1}^{(s)}.$$

- Calculate the innovation covariance:

$$\Sigma_{\mathcal{Y},k|k-1}^{(s)} = \sum_{j=1}^{P_s} w_j^{(s)}(Y_{j,k|k-1}^{(s)} - \hat{Y}_{k|k-1}^{(s)})(Y_{j,k|k-1}^{(s)} - \hat{Y}_{k|k-1}^{(s)})^T + R_k$$

- Calculate the cross-covariance:

$$\Sigma_{\mathcal{XY},k|k-1}^{(s)} = \sum_{j=1}^{P_s} w_j^{(s)}(\tilde{\mathcal{X}}_{j,k|k-1}^{(s)} - \hat{\mathcal{X}}_{k|k-1}^{(s)})(Y_{j,k|k-1}^{(s)} - \hat{Y}_{k|k-1}^{(s)})^T$$

- Calculate the Kalman gain:

$$K_k^{(s)} = \Sigma_{\mathcal{XY},k|k-1}^{(s)}\Sigma_{\mathcal{Y},k|k-1}^{(s)-1}$$

- Compute the posterior mean:

$$\hat{\mathcal{X}}_{k|k}^{(s)} = \hat{\mathcal{X}}_{k|k-1}^{(s)} + K_k^{(s)}(\mathcal{Y}_k - \hat{Y}_{k|k-1}^{(s)})$$

- Compute the posterior covariance:

$$\Sigma_{\mathcal{X},k|k}^{(s)} = \Sigma_{\mathcal{X},k|k-1}^{(s)} - K_k^{(s)}\Sigma_{\mathcal{Y},k|k-1}^{(s)}K_k^{(s)T}$$

5.6.4 Simulation example

The MSGHF algorithm is used for the estimation of amplitude and frequency of a multi-harmonic signal usually encountered in many engineering fields such as communications [31, 116], power systems [36], etc. As the magnitude and the frequency are assumed to be constant with some variations, the discrete state-space is considered to follow a random walk evolution. So the state-space is represented by

$$\mathcal{X}_k = A\mathcal{X}_{k-1} + \eta_k, \tag{5.36}$$

where $A = I_{3\times 3}$ and $\eta_k \sim \mathcal{N}(0,Q)$. This defines the state variables, $\mathcal{X} = [f_1, f_2, f_3, a_1, a_2, a_3]^T$, where f_i and a_i are the frequency and amplitude of the ith sinusoid. The process noise covariance is defined as $Q = diag([\sigma_f^2 \ \sigma_f^2 \ \sigma_f^2 \ \sigma_a^2 \ \sigma_a^2 \ \sigma_a^2])$, where $\sigma_f^2 = 0.1\mu Hz^2/ms^2$ and $\sigma_a^2 = 0.5\mu V^2/ms^2$. The measurement equation under study can be represented as

$$y_k = \left[\begin{array}{c} \sum_{i=1}^{3} a_{i,k} cos(2\pi f_{i,k}k) \\ \sum_{i=1}^{3} a_{i,k} sin(2\pi f_{i,k}k) \end{array} \right] + v_k, \tag{5.37}$$

where $v_k \sim \mathcal{N}(0,R)$. Measurement noise covariance is $R = \sigma_n^2 I_{2\times 2}$, where $\sigma_n^2 = 0.09V^2$. Sampling frequency of the signal is set to be $5000Hz$.

The filter was initialized with initial estimate, $\hat{\mathcal{X}}^{(s)}$, which follows normal distribution with mean $\mathcal{X}_0^{(s)}$ and covariance $\Sigma_0^{(s)}$ i.e., $\hat{\mathcal{X}}^{(s)} \sim \mathcal{N}(\mathcal{X}_0^{(s)}, \Sigma_0^{(s)})$, where $\mathcal{X}_0 = [100 \ 1000 \ 2000 \ 5 \ 4 \ 3]^T$ and $\Sigma_0 = diag([200^2 \ 200^2 \ 200^2 \ 4 \ 4 \ 4])$. Both $\mathcal{X}_0$ and Σ_0 have to be partitioned accordingly. The states were estimated for 350 steps and 5000 Monte Carlo runs were done for confidence in the interpretation of the results. Validation of the results was done by comparing the root mean square error (RMSE) of the frequencies. The RMSE of frequencies for the k-th step can be defined as

$$RMSE_{f_k} = \sqrt{\frac{(MSE_{f_{1,k}} + MSE_{f_{2,k}} + MSE_{f_{3,k}})}{3}}, \tag{5.38}$$

where MSE is the mean square error for M number of Monte Carlo runs, expressed as $MSE_{i,k} = \frac{1}{M}\sum_{j=1}^{M}(\mathcal{X}_{i,j,k} - \hat{\mathcal{X}}_{i,k,j}^{(s)})^2$.

The number of univariate quadrature points per dimension for the GHF algorithm was set to be $\alpha = 4$ and the accuracy level of SGHF was set as $L = 3$. The condition of track-loss for each frequency was defined as, $\mathcal{X}_{end} - \hat{\mathcal{X}}_{end} < 100Hz$, where $\mathcal{X}_{end}$ and $\hat{\mathcal{X}}_{end}$ denotes the truth value and the estimated value of the state at the N-th step, respectively.

Performance was evaluated by plotting the root mean square error (RMSE) as described in Eq. (5.38). The resulting RMSE plot is shown in Figure 5.6. State partitions that were considered and the corresponding quadrature points required to be generated are shown in Table 5.1. From the plot, it can be clearly inferred that as the number of partitions increases, error in the estimation also increases slightly. Table 5.2 shows the performance with respect to track loss. From the table, we see that the track loss is higher when the state-space partition is more, again proving the point that estimation error will be higher for more state-space partitions.

From the above discussion, it is clear that the main advantage of multiple quadrature Kalman filtering [127] is to bypass the the curse of dimensionality problem and lower the quadrature points requirement without much compromising with the estimation accuracy.

TABLE 5.1: Different subspace partitioning and the number of quadrature points used by various filters

State partitioning	Filter	No. of quadrature points	Relative execution time
Full state-space	GHF	4096	1
	SGHF	97	0.026
$(f_1, f_2, f_3)(a_1, a_2, a_3)$	MQKF-S1	$64 \times 2 = 128$	0.031
	MSGHF-S1	$31 \times 2 = 62$	0.017
$(f_1, a_1)(f_2, a_2)(f_3, a_3)$	MQKF-S2	$16 \times 3 = 48$	0.013
	MSGHF-S2	$17 \times 3 = 51$	0.014
$(f_1)(f_2)(f_3)(a_1)(a_2)(a_3)$	MQKF-S3	$4 \times 6 = 24$	0.011
	MSGHF-S3	$5 \times 6 = 30$	0.012

A caveat regarding the limitation of MSGHF [127] results is in order. The difference between this technique and the conventional filtering technique is that here each step is performed in parallel by the bank of filters. But, all filters are aware of the estimates delivered by the rest of the filters. Assumptions made in this scheme are 1) cross-terms between each pair of subspaces are neglected 2) coupling among the filters is achieved with the point estimates. State partitioning does not imply that the filtering problem deals with the independent estimation problem, because the observations may contain the full state information. Since possible process noise correlation among subspaces are not

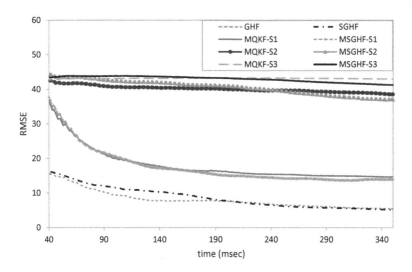

FIGURE 5.6: Performance of MSGHF for different state partitions [127].

TABLE 5.2: Track loss percentage

Freq.	GHF	SGHF	MQKF-S1	MSGHF-S1	MQKF-S2	MSGHF-S2	MQKF-S3	MSGHF-S3
f_1	0.5%	1%	8.8%	8.6%	30.4%	30.5%	33.6%	33.6%
f_2	0%	0%	26.8%	26.6%	63.2%	62.7%	54.7%	54.8%
f_3	1%	1.5%	44.8%	44.8%	70.4%	67.4%	65%	65.8%

considered, there is a possibility of losing information. This leads to an error in estimation and possibly a major drawback in state partitioning technique. The partitioning technique should be guided by the application, the degree of nonlinearity and the computational resources available. The best way of partitioning is combining the states which are highly correlated and putting the uncorrelated ones in different subspaces. Finally depending on process and measurement equations, it may happen that state partitioning is not possible for a particular problem.

5.7   Summary

In this chapter we discussed the Gauss-Hermite quadrature rule of numerical integration. The rule is further applied to nonlinear filtering. Unfortunately, the Gauss-Hermite filter suffers from the curse of dimensionality problem, which means that the number of points required increases exponentially with dimension. A variant of the Gauss-Hermite filter has been discussed which is free from the curse of dimensionality problem. Further, the state partitioning technique is discussed along with the GHF which can considerably decrease its computational burden for many systems.

Chapter 6

Gaussian sum filters

6.1 Introduction 117
6.2 Gaussian sum approximation 118
    6.2.1 Theoretical foundation 118
    6.2.2 Implementation 120
    6.2.3 Multidimensional systems 121
6.3 Gaussian sum filter 122
    6.3.1 Time update 122
    6.3.2 Measurement update 123
6.4 Adaptive Gaussian sum filtering 124
6.5 Simulation results 125
    6.5.1 Problem 1: Single dimensional nonlinear system 125
    6.5.2 RADAR target tracking problem 129
    6.5.3 Estimation of harmonics 133
6.6 Summary 136

6.1 Introduction

The problem of state estimation of a nonlinear stochastic system from noisy measurements is considered throughout the book. It is inferred that the knowledge about posterior density of the state, which is to be estimated, conditioned on measurements gives a complete description regarding the system dynamics. When the systems are linear and the concerned noise pdfs are assumed to be Gaussian, the closed form solution of the posterior pdf (which remains Gaussian and characterized using the mean and covariance) exists in the minimum mean square error (MMSE) sense, which is known as the Kalman filter. However, the Bayesian recursion relations (Eq. (1.29) and (1.31)) cannot generally be solved in closed-form when the system is either nonlinear or non-Gaussian or both.

In the previous chapters, we approximate the a priori and the a posteriori pdfs as Gaussian, represented with deterministic sample points and weights and consider the mean as a point estimate. It is a fact that when a Gaussian distribution is subjected to a nonlinear transformation, the end result will no longer be Gaussian. So the assumption of Gaussian posteriori density leads to a wrn the distributioong estimate especially when becomes multi modal. In this chapter we shall relax the Gaussian assumption, and a solution to

117

obtain a priori and a posteriori pdf (approximately) for nonlinear and non-Gaussian system will be discussed. To do so, a concept will be introduced which approximates any arbitrary pdf, either continuous or having a finite number of discontinuities, as the convex combination of the Gaussian density functions. As the number of terms in the Gaussian sum increases without bound, the approximation converges uniformly to the actual density function. Further, any finite sum will become a valid density function.

6.2   Gaussian sum approximation

The main objective is to approximate an arbitrary a posteriori density function and its moments. The concept of approximating an arbitrary pdf with a weighted sum of Gaussian was first introduced by Sorenson and Alspach in [151]. Before that the pdfs were approximated using Edgeworth and Gram-Charlier expansions [152]. Edgeworth and Gram-Charlier expansions had a disadvantage that when the series are truncated substantially, the resulting density approximation may become negative leading the approximation to be an invalid density function. To avoid this, sometimes it is necessary to retain a large number of terms in the series which makes the approximation computationally unattractive. In Gaussian sum approximation that limitation is not present. More specifically the approximation is always a valid density function. Further, the sum converges to a valid density function as the number of Gaussian components increases.

6.2.1   Theoretical foundation

Let us consider a probability density function $p(x)$ which is required to be approximated. Here, we consider a scalar random variable x for simplicity of exposition. However, the generalization to a multidimensional case is not difficult and we shall do so once we understand the concept. Now let us further assume that the pdf $p(x)$ has the following properties.

(i) p is continuous at all but a finite number of locations.

(ii) $p(x) \geq 0$ for all x.

(iii) $\int_{-\infty}^{\infty} p(x)dx = 1$.

There exist families of functions which converge to impulse function when a parameter characterizing the family of a function tends to a limiting value. Let us denote them with $\delta_\lambda(x)$ and they satisfy the following conditions.

(i) $\delta_\lambda(x)$ is defined over $(-\infty, \infty)$ and is integrable over every interval.

(ii) $\delta_\lambda(x) \geq 0$ for all x and λ.

(iii) There exist λ_0, a such that $\lim\limits_{\lambda \to \lambda_0} \int_{-a}^{a} \delta_\lambda(x)dx = 1$

(iv) For every constant $\gamma > 0$, and $\gamma \leq |x| < \infty$; as $\lambda \to \lambda_0$, $\delta_\lambda \to 0$ uniformly.

As an example, the Normal distribution $\mathcal{N}_\lambda(x; 0, \lambda)$ forms a positive delta family function because as the variance $\lambda \to 0$, the Normal distribution, $\mathcal{N}(.)$ becomes a delta function or impulse function.

Now let us state an important result. The convolution of δ_λ and p i.e.,

$$\int_{-\infty}^{\infty} \delta_\lambda(x - u)p(u)du = p_\lambda(x) \tag{6.1}$$

converges uniformly to $p(x)$ on every interior subinterval of $(-\infty, \infty)$ [151]. If δ_λ satisfies the condition $\int_{-\infty}^{\infty} \delta_\lambda(x)dx = 1$, p_λ is a probability density function for all λ. As we have seen that the Gaussian density forms a delta function family, we write

$$\delta_\lambda(x) \triangleq \mathcal{N}_\lambda(x) = (2\pi\lambda^2)^{-\frac{1}{2}} exp[-\frac{1}{2}x^2/\lambda^2], \tag{6.2}$$

and the pdf, $p(x)$, is approximated as $p_\lambda(x)$ where

$$p_\lambda(x) = \int_{-\infty}^{\infty} p(u)\mathcal{N}_\lambda(x - u)du. \tag{6.3}$$

The above equation is very important as it provides the basis of Gaussian sum approximation.

The above equation can be approximated on any finite interval by a Riemann sum in the following way. We divided a bounded interval (a, b) into n subintervals such that $a = \xi_0 < \xi_1 < \cdots < \xi_n = b$. So an approximation of p_λ over the same interval becomes

$$p_{n,\lambda}(x) \approx \frac{1}{k} \sum_{i=1}^{n} p(x_i)\mathcal{N}_\lambda(x - x_i)[\xi_i - \xi_{i-1}]. \tag{6.4}$$

In each subinterval x_i is a point such that

$$p(x_i)[\xi_i - \xi_{i-1}] = \int_{\xi_{i-1}}^{\xi_i} p(x)dx. \tag{6.5}$$

The choice of x_i is possible by the mean-value theorem. k in Eq. (6.4) is a normalization constant such that $p_{n,\lambda}(x)$ becomes a pdf. So we can write

$$\frac{1}{k} \sum_{i=1}^{n} p(x_i)[\xi_i - \xi_{i-1}] = 1. \tag{6.6}$$

It can also be noted that for $(b-a)$ sufficiently large, k can be made arbitrarily close to 1. From the above discussion, it is clear that Eq. (6.4) represents an approximation of $p(x)$ with a sum of Gaussian pdfs.

6.2.2   Implementation

From the previous subsection, we have seen that a pdf that may have a finite number of discontinuities can be approximated by the Gaussian sum $p_{n,\lambda}$ as defined in Eq. (6.4). In practical cases, $p(x)$ should be approximated with an acceptable accuracy by considering a small number of terms of the series. Further, Gaussian pdfs with finite variance need to be taken for summation in practice. It is convenient and practical to write the Gaussian sum approximation (Eq. (6.4)) as

$$\hat{p}_n(x) = \sum_{i=1}^{n} \alpha_i \mathcal{N}_{\sigma_i}(x - \mu_i), \tag{6.7}$$

where

$$\sum_{i=1}^{n} \alpha_i = 1; \alpha_i \geq 0, \forall i. \tag{6.8}$$

Please note that unlike the Eq. (6.4), each term in the above equation has a different variance and this enhances the flexibility for approximations using a finite number of terms. Further, as the number of Gaussian terms increases, the variance σ_i tends to become equal and negligibly small.

To approximate the $p(x)$ with Eq. (6.7) we need to determine $\alpha_i, \mu_i, \sigma_i$ so that we receive the best approximation. This can be achieved by minimizing the distance (norm) between $p(x)$ and $p_n(x)$ given by,

$$\| p(x) - \hat{p}_n(x) \| = \int_{-\infty}^{\infty} | p(x) - \sum_{i=1}^{n} \alpha_i \mathcal{N}_{\sigma_i}(x - \mu_i) | \, dx. \tag{6.9}$$

The above equation describes 1-norm while other higher norms can also be used as cost functions to obtain $\alpha_i, \mu_i, \sigma_i$. In many applications, calculation of moments becomes important; in such cases moments (generally first and second order) of the series are calculated and the true pdf is approximated with them. It should be kept in mind that the densities which have discontinuities generally are more difficult to approximate than are continuous functions [151].

Here, we take one example from [151]. Consider the following uniform density function

$$p(x) = \begin{cases} 1/4 & \text{for } -2 \leq x \leq 2, \\ 0 & \text{elsewhere} \end{cases}$$

which has zero mean and variance of $\frac{4}{3}$. This uniform density function has been approximated with the sum of Gaussian pdfs where μ_is of each Gaussian are equally spaced on $(-2, 2)$. The weighting factors α_i are normalized i.e., $\sum_{i=1}^{n} \alpha_i = 1$. The variances σ_i^2 are selected so that the cost in Eq. (6.9) is minimized. The approximation of the said uniform density function is made with the help of 6, 10 and 20 Gaussian components and shown in Figure 6.1 which is adapted from [151]. The accuracy of the approximation improves as the number of Gaussian components increases.

6.2.3 Multidimensional systems

The above discussion in the context of a single random variable can easily be extended for a multi random variable case. In the present context, for a stochastic system the state vector becomes a random variable and a posterior density function of state is to be reconstructed. For a random dynamic system, Eq. (6.7), the a posteriori probability density function, can be written as

$$p(\mathcal{X}_k|\mathcal{Y}_k) = \sum_{i=1}^{N_c} \bar{w}_{k|k}^i \mathcal{N}(\mathcal{X}_k; \hat{\mathcal{X}}_{k|k}^i, \Sigma_{k|k}^i), \tag{6.10}$$

where the weighted sum of N_c number of n dimensional Gaussian densities is considered. Like α_i, here the weights $\bar{w}_{k|k}$ also need to be normalized, i.e.,

$$\sum_{i=1}^{N_c} \bar{w}_{k|k}^i = 1. \tag{6.11}$$

Further, all weights must be positive, i.e.,

$$\bar{w}_{k|k}^i \geq 0. \tag{6.12}$$

If the posterior pdf is represent as above, the overall posterior mean of the states can readily be calculated as

$$\hat{\mathcal{X}}_{k|k} = \mathbb{E}[\mathcal{X}_k|\mathcal{Y}_k] = \sum_{i=1}^{N_c} \bar{w}_{k|k}^i \hat{\mathcal{X}}_{k|k}^i. \tag{6.13}$$

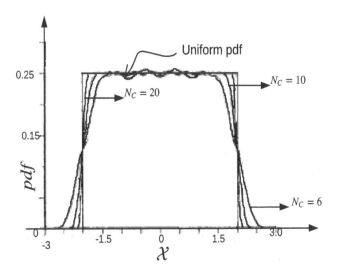

FIGURE 6.1: Gaussian sum approximation of uniform density function (adapted from [151]).

Similarly, the posterior error covariance can be calculated as

$$\Sigma_{k|k} = \mathbb{E}[(\mathcal{X}_k - \hat{\mathcal{X}}_{k|k})(\mathcal{X}_k - \hat{\mathcal{X}}_{k|k})^T]$$
$$= \sum_{i=1}^{N_c} \bar{w}_{k|k}^i \{\Sigma_{k|k}^i + (\hat{\mathcal{X}}_{k|k} - \hat{\mathcal{X}}_{k|k}^i)(\hat{\mathcal{X}}_{k|k} - \hat{\mathcal{X}}_{k|k}^i)^T\}. \qquad (6.14)$$

Simple formulas to calculate overall mean and covariance may be considered as an advantage of the Gaussian sum approach [2].

6.3   Gaussian sum filter

6.3.1   Time update

With the knowledge of density $p(\mathcal{X}_k|\mathcal{Y}_k)$ expressed as the sum of Gaussian terms and knowledge of the dynamic equations or the process model, it is possible to calculate approximately the one-step ahead prediction density $p(\mathcal{X}_{k+1}|\mathcal{Y}_k)$ also expressed as a sum of Gaussian pdfs. The previous step posterior density, $p(\mathcal{X}_k|\mathcal{Y}_k)$ is expressed as a sum of Gaussian pdfs as

$$\hat{p}(\mathcal{X}_k|\mathcal{Y}_k) = \sum_{i=1}^{N_c} \bar{w}_{k|k}^i \mathcal{N}(\mathcal{X}_k; \hat{\mathcal{X}}_{k|k}^i, \Sigma_{k|k}^i), \qquad (6.15)$$

and the extended Kalman filter theory is applied to yield an approximate expression for the one step prediction of each Gaussian distribution, $\mathcal{N}(\mathcal{X}_k; \hat{\mathcal{X}}_{k|k}^i, \Sigma_{k|k}^i)$ to another Gaussian distribution $\mathcal{N}(\mathcal{X}_{k+1}; \hat{\mathcal{X}}_{k+1|k}^i, \Sigma_{k+1|k}^i)$, where

$$\hat{\mathcal{X}}_{k+1|k}^i = \phi_k(\hat{\mathcal{X}}_{k|k}^i), \qquad (6.16)$$

$$\Sigma_{k+1|k}^i = A_k \Sigma_{k|k}^i A_k^T + Q_k, \qquad (6.17)$$

and

$$A_k = \frac{\partial \phi_k(\mathcal{X})}{\partial \mathcal{X}}\Big|_{\mathcal{X}=\hat{\mathcal{X}}_{k|k}^i}. \qquad (6.18)$$

In the time update the weight remains the same, i.e.,

$$\bar{w}_{k+1|k}^i = \bar{w}_{k|k}^i. \qquad (6.19)$$

The prior pdf becomes a series sum of Gaussian densities, i.e.,

$$\hat{p}(\mathcal{X}_{k+1}|\mathcal{Y}_k) = \sum_{i=1}^{N_c} \bar{w}_{k+1|k}^i \mathcal{N}(\mathcal{X}_{k+1}; \hat{\mathcal{X}}_{k+1|k}^i, \Sigma_{k+1|k}^i), \qquad (6.20)$$

and the overall prior mean becomes

$$\hat{\mathcal{X}}_{k+1|k} = \sum_{i=1}^{N_c} \bar{w}_{k+1|k}^i \hat{\mathcal{X}}_{k+1|k}^i. \tag{6.21}$$

The Gaussian sum approximation for the conditional state pdf approaches the true conditional pdf under the assumption that there is a sufficient number of the Gaussian kernels and the covariance of all Gaussian kernels is small enough.

6.3.2 Measurement update

In the previous subsection, we express the prior, $p(\mathcal{X}_{k+1}|\mathcal{Y}_k)$, as a weighted sum of Gaussian density and now we shall show that $p(\mathcal{X}_{k+1}|\mathcal{Y}_{k+1})$ can similarly be expressed when a new measurement $\mathcal{Y}_{k+1}$ becomes available. We use the extended Kalman filter for the measurement update of each component and after the measurement update the density becomes $\mathcal{N}(\mathcal{X}_k; \hat{\mathcal{X}}_{k|k}^i, \Sigma_{k|k}^i)$ where

$$\hat{\mathcal{X}}_{k+1|k+1}^i = \hat{\mathcal{X}}_{k+1|k}^i + K_{k+1}^i [\mathcal{Y}_{k+1} - \gamma_{k+1}(\hat{\mathcal{X}}_{k+1|k}^i)] \tag{6.22}$$

$$\Sigma_{k+1|k+1}^i = (I - K_{k+1}^i C_{k+1}^i)\Sigma_{k+1|k}^i \tag{6.23}$$

$$\Omega_{k+1}^i = C_{k+1}^i \Sigma_{k+1|k}^i C_{k+1}^{i^T} + R_{k+1} \tag{6.24}$$

$$K_{k+1}^i = \Sigma_{k+1|k}^i C_{k+1}^{i^T} \Omega_{k+1}^{i^{-1}} \tag{6.25}$$

$$C_{k+1}^i = \frac{\partial \gamma_{k+1}(\mathcal{X})}{\partial \mathcal{X}}\Big|_{\mathcal{X}=\hat{\mathcal{X}}_{k+1|k}^i}. \tag{6.26}$$

The new weights $\bar{w}_{k+1|k+1}^i$ are expressed as [2]

$$\bar{w}_{k+1|k+1}^i = \frac{\bar{w}_{k+1|k}^i \mathcal{N}(\mathcal{Y}_{k+1} - \gamma_{k+1}(\hat{\mathcal{X}}_{k+1|k}^i), \Omega_{k+1}^i)}{\sum_{i=1}^{N_c} \bar{w}_{k+1|k}^i \mathcal{N}(\mathcal{Y}_{k+1} - \gamma_{k+1}(\hat{\mathcal{X}}_{k+1|k}^i), \Omega_{k+1}^i)}. \tag{6.27}$$

Finally, the posterior distribution is the weighted sum of all the Gaussian pdfs, i.e.,

$$\hat{p}(\mathcal{X}_{k+1}|\mathcal{Y}_{k+1}) = \sum_{i=1}^{N_c} \bar{w}_{k+1|k+1}^i \mathcal{N}(\mathcal{X}_{k+1}; \hat{\mathcal{X}}_{k+1|k+1}^i, \Sigma_{k+1|k+1}^i). \tag{6.28}$$

The point estimate, which is the mean of the approximated pdf becomes

$$\hat{\mathcal{X}}_{k+1|k+1} = \sum_{i=1}^{N_c} \bar{w}_{k+1|k+1}^i \hat{\mathcal{X}}_{k+1|k+1}^i. \tag{6.29}$$

Here, for the time and measurement update of each Gaussian component, the formulas of the extended Kalman filter are used. Apart from using the extended Kalman filter, both the prior and posterior mean and covariance of each component can be determined using any quadrature filter *viz.* the unscented Kalman filter [157], cubature Kalman filter [100], Gauss-Hermite filter etc. We call them a proposal of the Gaussian sum filter. Further the PF can also be used as a proposal of the Gaussian sum filter [95].

6.4   Adaptive Gaussian sum filtering

In the earlier discussion, it was assumed that the weights of the Gaussian components do not change during the process update, see Eq. (6.19). According to [167], the assumption is only appropriate if the underlying system dynamics is linear or the system is at worst marginally nonlinear. In practice, this assumption is frequently violated resulting in a poor approximation of the conditional pdf [167]. In this section, a methodology is presented [167] to update Gaussian kernel weights during the propagation of state pdf between two measurements.

The weight update (during time update) is obtained by minimizing the integral squared error between the true pdf, $p(\mathcal{X}_{k+1}|\mathcal{Y}_k)$ and its Gaussian sum approximation $\hat{p}(\mathcal{X}_{k+1}|\mathcal{Y}_k)$. So our cost function becomes

$$J = \min_{\bar{w}^i_{k+1|k}} \frac{1}{2} \int |p(\mathcal{X}_{k+1}|\mathcal{Y}_k) - \hat{p}(\mathcal{X}_{k+1}|\mathcal{Y}_k)|^2 d\mathcal{X}_{k+1},$$

$$= \min_{\bar{w}^i_{k+1|k}} \frac{1}{2} \int |p(\mathcal{X}_{k+1}|\mathcal{Y}_k) - \sum_{i=1}^{N_c} \bar{w}^i_{k+1|k} \mathcal{N}(\mathcal{X}_{k+1}; \hat{\mathcal{X}}^i_{k+1|k}, \Sigma^i_{k+1|k})|^2 d\mathcal{X}_{k+1}.$$

The above cost function can be written as [167]

$$J = \frac{1}{2} \bar{W}^T_{k+1|k} \mathbb{M} W_{k+1|k} - \bar{W}^T_{k+1|k} \mathbb{N} \bar{W}_{k|k}, \tag{6.30}$$

where $\bar{W}_{k+1|k} = [\bar{w}^1_{k+1|k} \quad \bar{w}^2_{k+1|k} \cdots \bar{w}^{N_c}_{k+1|k}]^T$ is the prior weight vector. The symmetric matrix $\mathbb{M} \in \mathbb{R}^{N_c \times N_c}$ can be expressed as

$$\mathbb{M}_{i,j} = \int \mathcal{N}(\mathcal{X}_{k+1}; \hat{\mathcal{X}}^i_{k+1|k}, \Sigma^i_{k+1|k}) \mathcal{N}(\mathcal{X}_{k+1}; \hat{\mathcal{X}}^j_{k+1|k}, \Sigma^j_{k+1|k}) d\mathcal{X}_{k+1}$$

$$= |2\pi(\Sigma^i_{k+1|k} + \Sigma^j_{k+1|k})|^{-1/2} \exp[-\frac{1}{2}(\hat{\mathcal{X}}^i_{k+1|k} - \hat{\mathcal{X}}^j_{k+1|k})^T$$

$$(\Sigma^i_{k+1|k} + \Sigma^j_{k+1|k})^{-1}(\hat{\mathcal{X}}^i_{k+1|k} - \hat{\mathcal{X}}^j_{k+1|k})] \quad for \ i \neq j,$$

$$= |4\pi\Sigma^i_{k+1|k}|^{-1/2} \quad for \ i = j.$$

Matrix $\mathbb{N} \in \mathbb{R}^{N_c \times N_c}$ can be expressed as

$$\mathbb{N}_{ij} = \int \mathcal{N}(\phi(\mathcal{X}_{k|k}); \hat{\mathcal{X}}^j_{k+1|k}, \Sigma^j_{k+1|k}) \times \mathcal{N}(\mathcal{X}_k; \hat{\mathcal{X}}^i_{k|k}, \Sigma^i_{k|k}) d\mathcal{X}_k.$$

The elements in matrix $\mathbb{N}$ can be found using numerical methods. Here, cubature points and weights are used for this purpose.

Now, as the weights must be positive and to represent a pdf the sum of all weights must be unity, the constrained optimization problem can be described as

$$
\begin{aligned}
J = \min_{\bar{w}^i_{k+1|k}} & \frac{1}{2} \bar{W}^T_{k+1|k} \mathbb{M} \bar{W}_{k+1|k} - \bar{W}^T_{k+1|k} \mathbb{N} \bar{W}_{k|k} \\
s.t. & \sum_{i=1}^{N_c} \bar{w}^i_{k+1|k} = 1 \\
& \bar{w}^i_{k+1|k} \geq 0, \qquad i = 1, \cdots, N_c.
\end{aligned}
\tag{6.31}
$$

It is should be noted that Eq. (6.31) is a quadratic optimization problem and extremely efficient numerical methods exist to solve these problems. The derivation for obtaining $\mathbb{M}$ and $\mathbb{N}$ can be seen from [167].

It is also to be noted that for linear process models, matrices $\mathbb{M}$ and $\mathbb{N}$ become equivalent, and hence the weights are kept constant [167]. Hence there is no point in using the adaptive Gaussian sum approach to problems with the linear process model.

6.5 Simulation results

This section studies the performance of the Gaussian sum (GS) and adaptive Gaussian sum (AGS) filters where different sigma point filters are used as a proposal. During simulation, the initial probability density is split into Nc number of Gaussian probability density functions [44].

6.5.1 Problem 1: Single dimensional nonlinear system

A one-dimensional signal process is described with the following discretized process model [17]

$$\mathcal{X}_{k+1} = \phi(\mathcal{X}_k) + \eta_k, \tag{6.32}$$

where $\phi(\mathcal{X}_k) = \mathcal{X}_k + 5T\mathcal{X}_k(1 - \mathcal{X}_k^2)$, $\eta_k \sim \mathcal{N}(0, Q)$ and $Q = b^2T$. Here, b is a constant and T is sampling time. The measurement model is described using the relation

$$\mathcal{Y}_k = \gamma(\mathcal{X}_k) + v_k, \tag{6.33}$$

where $\gamma(\mathcal{X}_k) = T\mathcal{X}_k(1 - 0.5\mathcal{X}_k)$ and $R = d^2T$, d is a constant.

The process, mentioned here has three equilibrium points. They are at $+1$, -1 and 0. Here, the first two equilibrium points are stable and the other is unstable. Hence the state variable settles around any of the two stable equilibrium points. As the time progresses the state variable can switch from one equilibrium to another. Sampling time is taken as $T = 0.01s$ and b and d are taken as 0.5 and 0.1 respectively. Figure 6.2 shows the truth and Gaussian sum estimate of the state variable for a period of $4s$ when initial state value is taken as -0.2.

The above described problem has been solved using the Gaussian sum filter with the proposal of (i) UKF, (ii) CQKF, (iii) SGHF and (iv) $4n + 1$ points sigma point filter (as discussed in Section 3.4.2). The truth state is initialized as $\mathcal{X}_{0|0} = -0.2$ and the initial mean and covariance of filter are taken as $\hat{\mathcal{X}}_{0|0} = 0.8$ and $\Sigma_{0|0} = 2$ respectively. Here, two Gaussian components were considered, that is $Nc = 2$. Root mean square errors (RMSEs) are calculated from 1000 Monte-Carlo runs and plotted in Figure 6.3. From the plots it was observed that the Gaussian sum filter with the UKF and the NUKF proposals provide better results than the CQKF and SGHF proposals. A working MAT-LAB code of GS filter with CKF proposal for this problem has been listed in Listing 6.1.

```
1 % Estimation of state for a 1D problem
2 % For problem description: see S Bhaumik, Swati, "Cubature
      % quadrature Kalman filter", IET signal processing, vol 7,
      no 7, % 2013,
3 % CKF is taken as Gaussian sum proposal
4 % Program for Monte Carlo run
5 % Code credit: Rahul Radhakrishnan, IIT Patna
6 %
7 clc; clear all
```

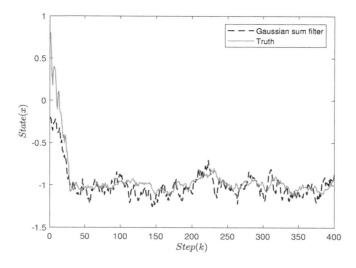

FIGURE 6.2: Problem 1: Truth value and Gaussian sum estimate of state.

```
 8 n=1; % system dimension
 9 % Values of constant parameters————————————————
10 kmax=400;   % value of $k_max$
11 M=100; % number of MC runs
12 b=0.5;d=0.1;T=0.01;
13 Q=b^2*T;R=d^2*T; % process and measurement noise parameters
14 d3=0;KP=.5;
15 randn('seed',0)  % Defining seed for random number generator if
       necessary
16 for mc=1:M        % Start of MC loop
17    % ———— Generation of truth————————————————
18    x(1)=-0.2;     % Initial states
19 for k=1:kmax
20    x(k+1)=x(k)+T*5*x(k)*(1-x(k)^2)+sqrt(Q)*randn(1);
21    y(k)=T*x(k)*(1-0.5*x(k))+sqrt(R)*randn(1);
22 end
23 % ———— Filtering step starts here————————————
24    % ———— Filter initialization————————————
25    xes(1)=0.8; P=5;
26    [COEFF,latent,explained] = pcacov(P);
27    XPG1=xes(:,1)+KP*sqrt(latent(1))*COEFF(:,1);
28    XPG2=xes(:,1)-KP*sqrt(latent(1))*COEFF(:,1);
29    XPG=[ XPG1  XPG2  ];
30    WT=[ 0.4 0.6];
31    PPSG1=P-KP^2*latent(1)*COEFF(:,1)*COEFF(:,1)';
32    PPSG2=PPSG1;PPSG=[PPSG1 PPSG2];
33    gc=2;   % No of Gaussian component
34
35    CP=sqrt(n)*[eye(n) -eye(n)];   % CKF points
36    for i=1:2*n, W(i)=1/(2*n);end % CKF wts
37    [row col]=size(W);
38 for k=1:kmax
39    PROBAB_SUM=0;
40    for gg=1:gc
41    S1=chol(PPSG(gg),'lower');
42      for i=1:col, chi(:,i)=XPG(gg)+S1*CP(:,i);end
43      for i=1:col, chinew(i)=chi(i)+T*5*chi(i)*(1-chi(i)^2);end
44    meanpred=0;
45  for i=1:col, meanpred=meanpred+W(i)*chinew(i);end % Prior mean
46    Pnew=0;
47    for i=1:col
48 Pnew=Pnew+(W(i)*(chinew(:,i)-meanpred)*(chinew(:,i)-meanpred)');
49    end
50    Pnew1=Pnew+Q;MEANpred(gg)=meanpred;PNEW(gg)=Pnew1;
51    end
52    for gg=1:gc
53       pp=chol(PNEW(gg),'lower');
54       for i=1:col,chii(:,i)=MEANpred(gg)+pp*CP(:,i);end
55       for i=1:col,znew(i)=T*chii(i)*(1-0.5*chii(i));end
56       zpred=0;
57       for i=1:col,zpred=zpred+W(i)*znew(:,i); end % Predicted
                 measurement
58       Pynew=0;
59          for i=1:col
60          Pynew=Pynew+W(i)*(znew(i)-zpred)*(znew(i)-zpred)';
61          end
62        Pyynew=Pynew+R;PXZnew=0;
```

```
63          for i=1:col
64              PXZnew=PXZnew+(W(i)*(chii(i)-MEANpred(gg))*(znew(i)-
                    zpred)'));
65          end
66          Kal_Gain=PXZnew/(Pyynew);
67          XPG(gg)=  MEANpred(gg)+ Kal_Gain*(y(k)-zpred);
68          PPSG(gg)=  PNEW(gg) - Kal_Gain* Pyynew*Kal_Gain';
69          PROB(gg) =1/sqrt(2*pi*sqrt(Pyynew))*exp(-(y(k)-zpred)
                    ^2/(Pyynew));
70          PROBAB_SUM=PROBAB_SUM + PROB(gg)*WT(gg);
71          WT_up(gg)=PROB(gg)*WT(gg); % Wt update during
                measurement update
72          end
73          WT1= WT_up/PROBAB_SUM;WT=WT1;xp_update=0;
74          for gg=1:gc,xp_update=xp_update+WT(gg)*XPG(:,gg); end %
                overall mean
75          xes(k+1)=xp_update; pp_update=0 ;
76          for gg=1:gc
77           pp_update= pp_update+  WT(gg)*(PPSG(gg)+...
78            ( XPG(gg)-xp_update)*( XPG(gg)-xp_update)'));
79           end
80          e(mc,k)=(xes(k)-x(k))^2;   %Squared error
81 end      % End of time loop
82 if abs(x(kmax)-xes(kmax))<1       % Cond for tracking
83          d3=d3+1;ee(d3,:)=e(mc,:);
84 end
85 end      % End of Monte Carlo loop
86          .
87 for k=1:kmax
88          e2=0;
89          for u=1:M, e2=e2+e(u,k);end
90          etsp(k)=(e2/M)^0.5;
91 end
92 figure(1), hold on, plot(etsp,'m')  % RMSE plot
93 figure(2), plot(x);hold on,plot(xes,'r') % Plot for single run
```

Listing 6.1: MATLAB code for GSF in the single dimensional nonlinear problem.

Another important parameter which accounts for filtering accuracy is the track-loss. A track is considered to be divergent if it moves away from its truth value without converging. Here, the track-loss condition was set as $|\mathcal{X}_{k_{max}} - \hat{\mathcal{X}}_{k_{max}}| > 1$, where k_{max} is the final time step which is 400 in this case. The percentage of track-loss incurred by the Gaussian sum filters, obtained from 1000 MC runs, and number of points are listed in Table 6.1.

The same problem has also been solved with the adaptive Gaussian sum filter (AGSF) with the same proposals as mentioned above. The RMSE of the filters from 1000 MC runs are calculated and plotted in Figure 6.4. From the figure it can be seen that AGSF is better than GSF and AGSF filters with all the mentioned proposals are performing with similar accuracy. Table 6.2 summarizes % track loss which is zero for different adaptive Gaussian sum filters.

TABLE 6.1: % track-loss and number of support points for Gaussian sum filter with different proposals.

Filter	Track-loss (%)	No. of points
GS-UKF	0.5	3
GS-CQKF	0.8	4
GS-SGHF	0.8	5
GS-NUKF	0.4	5

6.5.2 RADAR target tracking problem

Here, we consider a 7D, RADAR target tracking problem where the motive is to accurately track the trajectory of an airborne vehicle which maneuvers at a nearly constant speed and turn rate [13]. This kind of motion is referred to as nearly coordinated turn motion and is subjected to a small process noise. We solved a similar problem in Chapters 2 and 5.

The coordinated turn model is represented as $\mathcal{X}_{k+1} = F\mathcal{X}_k + \eta_k$. The state vector is $\mathcal{X} = [x \quad v_x \quad y \quad v_y \quad z \quad v_z \quad \Omega]^T$, where (x, y, z) are the positions,

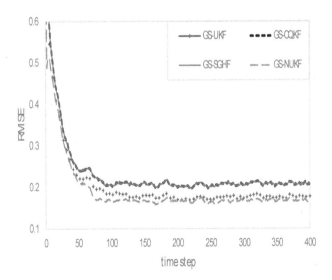

FIGURE 6.3: Problem 1: RMSE of Gaussian sum filters.

TABLE 6.2: % track-loss and number of support points for adaptive Gaussian sum filters.

Filter	Track-loss (%)	No. of points
AGS-UKF	0	3
AGS-CQKF	0	4
AGS-GHF	0	3
AGS-SGHF	0	5
AGS-NUKF	0	5

(v_x, v_y, v_z) are the velocities and Ω is the turn rate and

$$F = \begin{bmatrix} 1 & T & 0 & 0 & 0 & 0 & 0 \\ 0 & 1 & 0 & -T\Omega_k & 0 & 0 & 0 \\ 0 & 0 & 1 & T & 0 & 0 & 0 \\ 0 & T\Omega_k & 0 & 1 & 0 & 0 & 0 \\ 0 & 0 & 0 & 0 & 1 & T & 0 \\ 0 & 0 & 0 & 0 & 0 & 1 & 0 \\ 0 & 0 & 0 & 0 & 0 & 0 & 1 \end{bmatrix}.$$

The process noise is taken as white Gaussian with the covariance $Q = diag([0 \ \sigma_1^2 \ 0 \ \sigma_1^2 \ 0 \ \sigma_1^2 \ \sigma_2^2])$. Here, σ_1 is the standard deviation of velocities in all the 3 dimensions and σ_2 is the standard deviation of turn rate.

For this problem, the measurements received are assumed to be range, azimuth and elevation from a RADAR located at the origin. Hence the

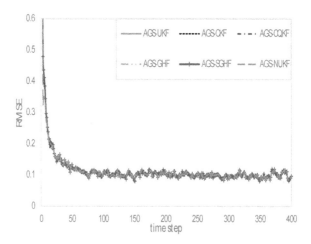

FIGURE 6.4: RMSE of adaptive Gaussian sum filters.

measurement equation is defined as

$$\mathcal{Y}_k = \begin{bmatrix} \sqrt{x_k^2 + y_k^2 + z_k^2} \\ \tan^{-1}(x_k/y_k) \\ \tan^{-1}(x_k/\sqrt{y_k^2 + z_k^2}) \end{bmatrix} + v_k.$$

The measurement noise v_k is zero mean white Gaussian with covariance $R = diag([\sigma_3^2, \sigma_4^2, \sigma_5^2])$, where σ_3, σ_4 and σ_5 are standard deviations of range, azimuth and elevation, respectively. Typical target trajectories for different turn rates have been plotted in Figure 6.5.

The above formulated problem has been solved with different Gaussian sum filters. The process and measurement noise covariance are defined with parameters $\sigma_1 = \sqrt{0.2}m/s$, $\sigma_2 = 7 \times 10^{-3}{}^\circ/s$, $\sigma_3 = 50m$, $\sigma_4 = 0.1^\circ$ and $\sigma_5 = 0.1^\circ$. The sampling time is taken as $T = 2s$ and the total observation is carried out for a period of $150s$. The initial estimates are assumed to be normally distributed with mean $\hat{\mathcal{X}}_{0|0}$ and covariance $\Sigma_{0|0}$, where $\hat{\mathcal{X}}_{0|0} = [1000m \ 0 \ 2650m \ 150m/s \ 200m/s \ 0 \ 0^\circ/s]^T$ and $\Sigma_{0|0} = diag([50m^2 \ 5(m/s)^2 \ 50m^2 \ 5(m/s)^2 \ 50m^2 \ 5(m/s)^2 \ 0.1(^\circ/s)^2])$. Here, three Gaussian components are used, i.e., $Nc = 3$.

The performance of the Gaussian sum filters with the proposals of UKF, CQKF, SGHF and NUKF is compared in terms of RMSE and track-loss count.

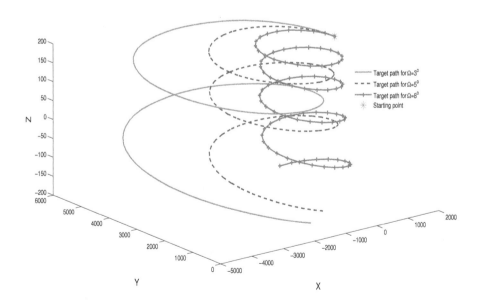

FIGURE 6.5: Target paths for different turn rates.

the RMSE of the radial position is calculated out of 500 Monte Carlo runs and plotted in Figure 6.6. From the figure it can be seen that all the filters perform with the same accuracy. From the RMSE plot of resultant velocity shown in Figure 6.7, a similar type of performance can be inferred, where all Gaussian sum filters performed with comparable accuracy. Similar performance is observed in Figure 6.8 where the RMSE of the turn rate (Ω) is plotted.

As mentioned earlier, track loss is counted for different Gaussian sum filters. A track is said to be divergent when the estimation error of the resultant position at the last time instant is greater than $50m$, i.e., $poserror_{k_{max}} \geq 50m$. Table 6.3 shows the percentage track-loss, number of points and computational load for all the filters.

The performance of adaptive Gaussian sum filters is also studied in terms of RMSE. Figures 6.9, 6.10 and 6.11 show the RMSE of the resultant position, velocity and turn rate obtained from 500 Monte Carlo runs. All the Gaussian sum filters show comparable performance.

The number of diverged tracks incurred by all the adaptive Gaussian sum filters are counted and they are the same as mentioned in Table 6.3. It should also be noted that for this problem, no significant improvement in performance is achieved by adaptive Gaussian sum filters (compared to Gaussian sum filters).

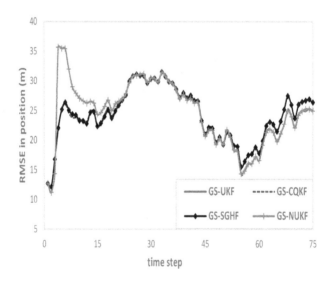

FIGURE 6.6: RMSE in position for Gaussian sum filters in a radar target tracking problem.

6.5.3 Estimation of harmonics

Recall the problem of amplitude and magnitude estimation of superimposed harmonics which has been described in Subsection 5.6.4. In this section,

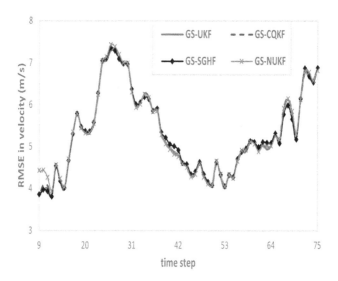

FIGURE 6.7: RMSE of velocity for Gaussian sum filters in a radar target tracking problem.

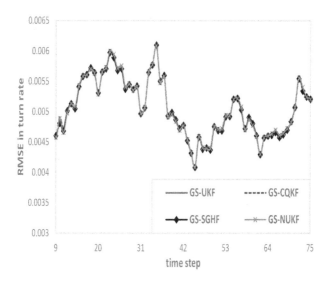

FIGURE 6.8: RMSE in turn rate for Gaussian sum filters in a radar target tracking problem.

we apply the Gaussian sum filters to the same problem. Here, we have not implemented the adaptive Gaussian sum filter because there is no advantage in doing so as the process is linear and the process noise is Gaussian. The

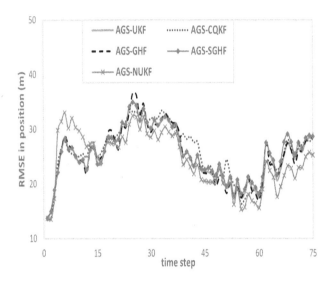

FIGURE 6.9: RMSE in position for adaptive Gaussian sum filters in a radar target tracking problem.

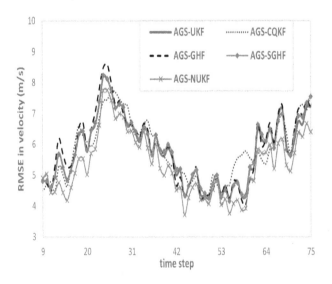

FIGURE 6.10: RMSE in velocity for adaptive Gaussian sum filters in a radar target tracking problem.

TABLE 6.3: % of track-loss, number of support points and relative computational time for Gaussian sum filters in a RADAR target tracking problem.

Filter	Track-loss (%)	No. of points	Rel. compu. time
GS-UKF	6	15	1
GS-CQKF	6	28	1.56
GS-SGHF	8	127	3.7
GS-NUKF	6	29	1.58

process and measurement models are considered same as that discussed in Subsection 5.6.4. Since there are three signals, a total of six states, three amplitudes and three frequencies have to be estimated. The error covariance for state variables f_i and a_i, are taken as $\sigma_{f_i}^2 = 0.1 \times 10^{-4}$ and $\sigma_{a_i}^2 = 0.5 \times 10^{-4}$, respectively for $i = 1, 2, 3$. To define the measurement noise covariance, the variance of individual measured signals is assumed to be $\sigma_n^2 = 0.09V^2$. The sampling frequency of the signal is set to be $5000Hz$ and $Nc = 3$.

The filters are initialized with initial estimate, $\hat{\mathcal{X}}_{0|0}$, which follows normal distribution with mean $\mathcal{X}_{0|0}$ and covariance $\Sigma_{0|0}$, *i.e*, $\hat{\mathcal{X}}_{0|0} \sim \mathcal{N}(\mathcal{X}_{0|0}, \Sigma_{0|0})$, where $\mathcal{X}_{0|0} = [100\ 1000\ 2000\ 5\ 4\ 3]^T$ and $\Sigma_{0|0} = diag([50^2\ 50^2\ 50^2\ 4\ 4\ 4])$. The states are estimated for 350 time steps and 1000 Monte Carlo runs. Validation of results was done by comparing the root mean square error (RMSE) of

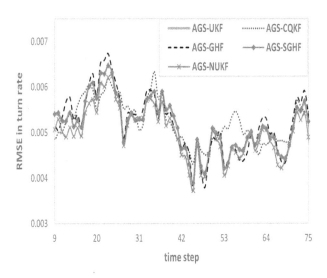

FIGURE 6.11: RMSE in turn rate for adaptive Gaussian sum filters in a radar target tracking problem.

the frequencies and amplitudes. The relation for calculating the RMSE of
frequency and amplitude is the same as mentioned in Eq. (5.38). RMSEs of
frequency and amplitude, calculated from 1000 Monte Carlo runs are plotted
in Figures 6.12 and 6.13 respectively. From these figures, it can be noted that
all the Gaussian sum filters work with similar accuracy.

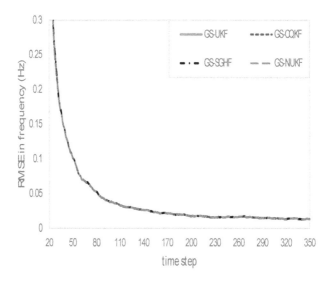

FIGURE 6.12: RMSE of frequency (Hz).

The number of the diverged track has also been counted. In the case of
frequencies, a track is considered divergent when the error between truth and
the estimate is above $5Hz$. For amplitude, a track divergence is defined when
the error is above $0.5V$. It has been found that none of the GS filters incurred
even a single diverged track.

6.6   Summary

This chapter describes the concept of representing an arbitrary probability
density function with the help of a weighted sum of many Gaussian distributions. This concept has been applied to state estimation problems which leads
to a nonlinear estimation method known as a Gaussian sum filter. Initially,
the extended Kalman filter is used to generate a single Gaussian component
which we call a proposal. Any nonlinear filtering heuristic can be used as a
proposal of the GSF. Here, we use different deterministic sigma point filters
to generate individual Gaussian components of GSF. Further, to account for

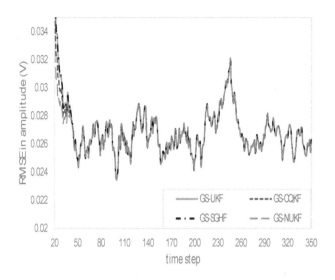

FIGURE 6.13: RMSE of amplitude in volt (V).

the possible approximation error while representing the prior pdf as a sum of many Gaussian densities in case of nonlinear process model, a weight update scheme is implemented. This scheme was termed as adaptive Gaussian sum filter. The performance of these filters was studied by implementing it on simple problems. The filtering performance is compared in terms of the RMSE, percentage of track-loss and relative computational time. From the results, it was observed that the adaptive Gaussian sum filters performed with similar or improved accuracy when compared to Gaussian sum filters.

Chapter 7

Quadrature filters with randomly delayed measurements

7.1	Introduction	139
7.2	Kalman filter for one step randomly delayed measurements	140
7.3	Nonlinear filters for one step randomly delayed measurements	143
	7.3.1 Assumptions	144
	7.3.2 Measurement noise estimation	144
	7.3.3 State estimation	145
7.4	Nonlinear filter for any arbitrary step randomly delayed measurement	146
	7.4.1 Algorithm	153
7.5	Simulation	154
7.6	Summary	155

7.1 Introduction

In earlier chapters, we have discussed a few filtering techniques for nonlinear systems. All the filters assume that the current measurement is available at every time instant without any delay. In practice, this is not always the case.

For example, in networked control systems, where the estimator is remotely located, delays and packet drops are inevitable during data transmission through unreliable communication channels. The estimator may receive the measurements with a delay in time. In such a case, the measurement received at the time instant 't' may actually belong to any time '$t-\tau$' where $0 \leq \tau < t$. The equality sign represents that the measurement may be non-delayed and belongs to the current time step itself. This delay may occur before the arrival of the measurement to the sensor or during its propagation from the sensor.

The delay that arises during transmission is generally not deterministic because it depends on the congestion in the network. It is random in nature and is described by stochastic parameters. Our objective of this chapter is to discuss deterministic sample point filters in the presence of random delay in measurement.

The estimation with delayed measurement was first solved for linear systems by Ray et al. [129] where one-step random delay is considered. The random delay has been modeled by Bernoulli trial with values zero or one which indicates the measurement arrived on time or delayed by one sampling time. The case was further solved for a nonlinear system for single step and two steps randomly delayed measurements.

Hermoso-Carazo et al. developed a nonlinear filtering algorithm for one-time step [67] and two-time step [68] randomly delayed measurements using the extended and the unscented Kalman filter. Later, Wang et al. [175] incorporated the cubature Kalman filter (CKF) to solve the nonlinear filtering problems with one-step randomly delayed measurement. The work has been extended for correlated noises in [176]. Recently, a one-step randomly delayed estimation problem for a nonlinear system has been solved with the Gauss-Hermite filter and its variant [143].

Further, the estimation problem has been solved for any arbitrary step randomly delayed measurement in linear systems. In this context the work of S. Sun and others must be acclaimed [163, 162, 160, 161]. Apart from those papers, there are other notable publications [182, 146, 105] which should also be mentioned in the present context.

The rest of the chapter is organized as follows. We discuss filtering in a linear system with a single step random delay in measurements in the next section. In Section 7.3, nonlinear filters with one step randomly delayed measurements are formulated. The next section deals with nonlinear estimation methods for any arbitrary step randomly delayed measurements. Finally the chapter ends with a brief summary.

7.2 Kalman filter for one step randomly delayed measurements

Our standard notation for linear systems is repeated here for the sake of completeness. A linear system with Gaussian noise in discrete time is expressed as

$$\mathcal{X}_{k+1} = A_k \mathcal{X}_k + \eta_k, \tag{7.1}$$

and

$$\mathcal{Y}_k = C_k \mathcal{X}_k + v_k, \tag{7.2}$$

where $\mathcal{X}_k \in \mathbb{R}^n$, denotes the state vector of a system and $\mathcal{Y}_k \in \mathbb{R}^p$ is the measurement at any instant k, where $k = \{0, 1, 2, 3, \cdots\}$. The sensor measurement is transmitted through an unreliable network where a single step delay may occur. It means that the measurement reaches either without delay or with a maximum one step delay. A single step randomly delayed measurement can be expressed as

$$\mathcal{Z}_k = (1 - \beta_k)\mathcal{Y}_k + \beta_k \mathcal{Y}_{k-1}, \tag{7.3}$$

where β is a Bernoulli random variable taking values either 0 or 1 with probability

$$\mathbb{P}(\beta = 1) = \mathbb{E}[\beta] = \mathrm{p},$$
$$\mathbb{P}(\beta = 0) = 1 - \mathrm{p}, \tag{7.4}$$
$$\text{and} \quad \mathbb{E}[(\beta - \mathrm{p})^2] = \mathrm{p}(1 - \mathrm{p}).$$

Here, we assume that there is no missing measurement and all the measurements reach either instantaneously or with a maximum delay of one time step. Our objective is to estimate states $(\hat{\mathcal{X}}_{k|k})$ from delayed measurement, $\mathcal{Z}_k$. Now let us assume the following:

(i) Plant noise, η_k is white Gaussian sequence with $\mathbb{E}[\eta_k] = 0$ and $\mathbb{E}[w_k w_l^T] = Q_k \delta_{kl}$.

(ii) Sensor noise, v_k is white Gaussian with $\mathbb{E}[v_k] = 0$ and $\mathbb{E}[v_k v_l^T] = R_k \delta_{kl}$.

(iii) Random sequences η_k, v_k, and β_k are mutually independent.

(iv) Measurement delays, $\beta_k \in [0, 1]$ are white, i.e., $\mathbb{P}(\beta_k \beta_l) = \mathbb{P}(\beta_k)\mathbb{P}(\beta_l)$ for $\forall k \neq l$.

(v) The initial state $\mathcal{X}_0$ is Gaussian and is statistically independent of the noises.

Under the above assumptions and initial condition, for a system described with Eqs. (7.1) - (7.3), a closed form solution of $\hat{\mathcal{X}}_{k|k}$ is available which we shall discuss now.

The posterior estimation error at any time step k is

$$e_{k|k} = \hat{\mathcal{X}}_{k|k} - \mathcal{X}_k. \tag{7.5}$$

The mean squared error (MSE) of an estimate is

$$J_k = \mathbb{E}[e_{k|k} e_{k|k}^T]. \tag{7.6}$$

Our objective is to determine $\hat{\mathcal{X}}_{k|k}$ so that $J_{k|k}$ is minimized. We assume that the estimator is in the form

$$\hat{\mathcal{X}}_{k|k} = \hat{\mathcal{X}}_{k|k-1} + K_k(\mathrm{H}_k \hat{\mathcal{X}}_{k|k-1} - \mathcal{Z}_k). \tag{7.7}$$

The above equation can further be written as

$$\hat{\mathcal{X}}_{k|k} = (I + K_k \mathrm{H}_k)\hat{\mathcal{X}}_{k|k-1} - K_k \mathcal{Z}_k,$$
$$= L_k \hat{\mathcal{X}}_{k|k-1} - K_k \mathcal{Z}_k, \tag{7.8}$$

where we denote $I + K_k \mathrm{H}_k = L_k$. We have to determine the Kalman gain, K_k and L_k. The error matrix can further be manipulated as

$$e_{k|k} = L_k \hat{\mathcal{X}}_{k|k-1} - K_k \mathcal{Z}_k - \mathcal{X}_k$$
$$= L_k(\mathcal{X}_k + e_{k|k-1}) - K_k[(1 - \beta)\mathcal{Y}_k + \beta \mathcal{Y}_{k-1}] - \mathcal{X}_k. \tag{7.9}$$

Substituting $\mathcal{Y}_k$ from the measurement equation and $\mathcal{X}_k$ from the process equation, we receive

$$
\begin{aligned}
&e_{k|k} \\
&= L_k(\mathcal{X}_k + e_{k|k-1}) - K_k[(1-\beta_k)(C_k\mathcal{X}_k + v_k) + \beta_k C_{k-1}\mathcal{X}_{k-1} + \beta_k v_{k-1}] - \mathcal{X}_k \\
&= L_k(A_{k-1}\mathcal{X}_{k-1} + \eta_{k-1} + e_{k|k-1}) - K_k[(1-\beta_k)(C_k A_{k-1}\mathcal{X}_{k-1} + C_k\eta_{k-1} + v_k) \\
&\quad + \beta_k C_{k-1}\mathcal{X}_{k-1} + \beta_k v_{k-1}] - A_{k-1}\mathcal{X}_{k-1} - \eta_{k-1},
\end{aligned}
$$
(7.10)

or,

$$
\begin{aligned}
e_{k|k} =&[-(1-\beta_k)K_k C_k A_{k-1} + L_k A_{k-1} - A_{k-1} - \beta_k K_k C_{k-1}]\mathcal{X}_{k-1} \\
&+ [-(1-\beta_k)K_k C_k + L_k - I_n]\eta_{k-1} \\
&+ L_k e_{k|k-1} - (1-\beta_k)K_k v_k - \beta_k K_k v_{k-1}.
\end{aligned}
$$
(7.11)

As the noises are zero mean and the estimator is assumed as unbiased, $\mathbb{E}[e_{k|k}] = 0$, or

$$
\mathbb{E}[-(1-\beta_k)K_k C_k A_{k-1} + L_k A_{k-1} - A_{k-1} - \beta_k K_k C_{k-1}]\mathbb{E}[\mathcal{X}_{k-1}] = 0. \quad (7.12)
$$

Since, in general, $\mathbb{E}[\mathcal{X}_{k-1}] \neq 0$,

$$
L_k = I_n + K_k[(1-\mathrm{p})C_k + \mathrm{p}C_{k-1}A_{k-1}^{-1}].
$$
(7.13)

The above equation expresses L_k in terms of Kalman gain K_k. As we stated earlier, the Kalman filter minimizes the mean squared error. So, the problem becomes an optimization problem as follows:

$$
\min_{K_k} J_k = \min_{K_k} \; \mathrm{Trace}(\mathbb{E}[e_{k|k}e_{k|k}^T|\mathcal{Z}_k]).
$$
(7.14)

After solving the optimization problem, the expression of Kalman gain becomes

$$
\begin{aligned}
K_k =&[(1-\mathrm{p})\Sigma_{k|k-1}C_k^T + \mathrm{p}A_{K-1}\Sigma_{k-1|k-1}C_{k-1}^T] \\
&\times \{[(1-\mathrm{p})C_k + \mathrm{p}C_{k-1}A_{K-1}^{-1}]\Sigma_{k|k-1}[(1-\mathrm{p})C_k + \mathrm{p}C_{k-1}A_{k-1}^{-1}]^T \\
&+ \mathrm{p}(1-\mathrm{p})[C_k - C_{k-1}A_{k-1}^{-1}]^T\mathbb{E}[\mathcal{X}_k\mathcal{X}_k^T|\mathcal{Z}_k][C_k - C_{k-1}A_{k-1}^{-1}]^T \\
&- \mathrm{p}C_{k-1}A_{k-1}^{-1}Q_{k-1}^T A_{k-1}^{-T}C_{k-1}^T + \mathrm{p}R_{k-1} + (1-\mathrm{p})R_k\}^{-1},
\end{aligned}
$$
(7.15)

where

$$
\mathbb{E}[\mathcal{X}_k\mathcal{X}_k^T|\mathcal{Z}_k] = \hat{\mathcal{X}}_{k|k-1}\hat{\mathcal{X}}_{k|k-1}^T + \Sigma_{k|k-1}.
$$
(7.16)

The above expression for Kalman gain is not derived here. To prove the expression, readers are requested to see [129]. The time update expression of the error covariance matrix is the same as the ordinary Kalman filter as mentioned below:

$$
\Sigma_{k|k-1} = A_{k-1}\Sigma_{k-1|k-1}A_{k-1}^T + Q_{k-1}.
$$
(7.17)

The measurement update of the error covariance matrix of the one step random delay Kalman filter is given by

$$
\begin{aligned}
\Sigma_{k|k} =& L_k \Sigma_{k|k-1} L_k^T + \mathrm{p}(1-\mathrm{p}) K_k [C_k - C_{k-1} A_{k-1}^{-1}] \mathbb{E}[\mathcal{X}_k \mathcal{X}_k^T | \mathcal{Z}_k] \\
& \times [C_k - C_{k-1} A_{k-1}^{-1}]^T K_k^T - \mathrm{p} K_k C_{k-1} A_{k-1}^{-1} Q_{k-1} A_{k-}^{-T} C_{k-1}^T K_k^T \\
& + \mathrm{p}(Q_{k-1} A_{k-1}^{-T} C_{k-1}^T K_k^T + K_k C_{k-1} A_{K-1}^{-1} Q_{k-1}) \\
& + (1-\mathrm{p}) K_k R_k K_k^T + \mathrm{p} K_k R_{k-1} K_k^T.
\end{aligned}
\tag{7.18}
$$

It should be noted that for non-delayed measurements, $\mathrm{p} = 1$ and for such a case the above equations reduce to standard Kalman filter equations.

7.3 Nonlinear filters for one step randomly delayed measurements

Let us recall the state space equations of a nonlinear system:

$$
\mathcal{X}_{k+1} = \phi_k(\mathcal{X}_k) + \eta_k,
\tag{7.19}
$$

and

$$
\mathcal{Y}_k = \gamma_k(\mathcal{X}_k) + v_k,
\tag{7.20}
$$

where $\mathcal{X}_k \in \mathbb{R}^n$, denotes the state vector of a system and $\mathcal{Y}_k \in \mathbb{R}^p$ is the measurement at any instant k where $k = \{0, 1, 2, \cdots, N\}$. $\phi_k(\cdot)$ and $\gamma_k(\cdot)$ are known nonlinear functions of $\mathcal{X}_k$. The process noise $\eta_k \in \mathbb{R}^n$ and measurement noise $v_k \in \mathbb{R}^p$ are assumed to be uncorrelated, white and distributed normally with covariance Q_k and R_k respectively.

The one step random delayed measurement equation is provided in Eq. (7.3). With the substitution of the measurement equation, the delayed measurement equation becomes

$$
\begin{aligned}
\mathcal{Z}_k &= (1 - \beta_k) \mathcal{Y}_k + \beta_k \mathcal{Y}_{k-1} \\
&= (1 - \beta_k)(\gamma_k(\mathcal{X}_k) + v_k) + \beta_k(\gamma_{k-1}(\mathcal{X}_{k-1}) + v_{k-1}).
\end{aligned}
\tag{7.21}
$$

For such a system, nonlinear estimators based on deterministic sample points are available in the literature. Wang et al. [176] modified the cubature Kalman filter (CKF) to solve the problem. Carazo et al. [68] [67] and Singh et al. [143] modified the unscented Kalman filter and the Gauss-Hermite Kalman filter to estimate $\mathcal{X}_{k|k}$. Whatever be the method, the algorithm remains the same except for the fact that the support points and weights change for each case.

At the beginning, the state vector is augmented with the measurement noise. So, the augmented state vector becomes $\mathcal{X}_k^a = [\mathcal{X}_k^T \quad v_k^T]^T$, and we have to estimate the posterior pdf of the augmented states, *viz.* $p(\mathcal{X}_k^a | \mathcal{Z}_k)$.

7.3.1 Assumptions

The probability densities at each time step in general are non-Gaussian in nature. The quadrature filters assume them as Gaussian and approximate with the first and second order moments. We assume the following:

- The probability density function of state, $p(\mathcal{X}_{k+1}|\mathcal{Z}_{k+1})$, is Gaussian with mean $\hat{\mathcal{X}}_{k+1|k+1}$, and covariance $\Sigma_{k+1|k+1}$.

- The probability density function of measurement noise, $p(v_{k+1}|\mathcal{Z}_{k+1})$ is Gaussian with mean $\hat{v}_{k+1|k+1}$ and covariance $\Sigma_{k+1|k+1}^{vv}$.

- The pdf of augmented states follows Gaussian distribution, i.e.,

$$p(\mathcal{X}_{k+1}^{a}|\mathcal{Z}_{k+1}) = \mathcal{N}(\mathcal{X}_{k+1}^{a}; \hat{\mathcal{X}}_{k+1|k+1}^{a}, \Sigma_{k+1|k+1}^{a})$$

where

$$\hat{\mathcal{X}}_{k+1|k+1}^{a} = \left[\begin{array}{c} \hat{\mathcal{X}}_{k+1|k+1} \\ \hat{v}_{k+1|k+1} \end{array} \right],$$

and

$$\Sigma_{k+1|k+1}^{a} = \left[\begin{array}{cc} \Sigma_{k+1|k+1} & \Sigma_{k+1|k+1}^{\mathcal{X}v} \\ (\Sigma_{k+1|k+1}^{\mathcal{X}v})^{T} & \Sigma_{k+1|k+1}^{vv} \end{array} \right],$$

with $\hat{v}_{k+1|k+1}$ being the posterior estimate for measurement noise, while $\Sigma_{k+1|k+1}^{vv}$ and $\Sigma_{k+1|k+1}^{\mathcal{X}v}$ are the noise covariance and cross-covariance between the state and measurement noise respectively.

- The one-step predictive pdf of $\mathcal{X}_{k+1}$ is Gaussian:

$$p(\mathcal{X}_{k+1}|\mathcal{Z}_{k}) = \mathcal{N}(\mathcal{X}_{k+1}; \hat{\mathcal{X}}_{k+1|k}, \Sigma_{k+1|k}). \tag{7.22}$$

- The one-step predictive pdf of delayed measurement is Gaussian, i.e.,

$$p(\mathcal{Z}_{k+1}|\mathcal{Z}_{k}) = \mathcal{N}(\mathcal{Z}_{k+1}; \hat{\mathcal{Z}}_{k+1|k}, \Sigma_{k+1|k}^{\mathcal{Z}\mathcal{Z}}). \tag{7.23}$$

- The pdfs of non-delayed measurements, $p(\mathcal{Y}_{k}|\mathcal{Z}_{k})$ and $p(\mathcal{Y}_{k+1}|\mathcal{Z}_{k})$ are Gaussian with mean $\hat{\mathcal{Y}}_{k|k}$ and $\hat{\mathcal{Y}}_{k+1|k}$, and covariance $\Sigma_{k|k}^{yy}$, and $\Sigma_{k+1|k}^{yy}$ respectively.

The estimation of the augmented states can be done in two steps [80], namely (i) measurement noise estimation (ii) state estimation.

7.3.2 Measurement noise estimation

The first and the second order moments of the prior and the posterior pdf of measurements are given by

$$\hat{\mathcal{Y}}_{k+1|k} = \int \gamma_{k+1}(\mathcal{X}_{k+1}) \mathcal{N}(\mathcal{X}_{k+1}; \hat{\mathcal{X}}_{k+1|k}, \Sigma_{k+1|k}) d\mathcal{X}_{k+1} \tag{7.24}$$

$$\Sigma^{yy}_{k+1|k} = \int \gamma_{k+1}(\mathcal{X}_{k+1})\gamma^T_{k+1}(\mathcal{X}_{k+1})\mathcal{N}(\mathcal{X}_{k+1}; \hat{\mathcal{X}}_{k+1|k}, \Sigma_{k+1|k})d\mathcal{X}_{k+1}$$
$$- \hat{\mathcal{Y}}_{k+1|k}\hat{\mathcal{Y}}^T_{k+1|k} + R_{k+1} \tag{7.25}$$

$$\hat{\mathcal{Y}}_{k|k} = \int [\gamma_k(\mathcal{X}_k) + v_k]\mathcal{N}(\mathcal{X}^a_k; \hat{\mathcal{X}}^a_{k|k}, \Sigma^a_{k|k})d\mathcal{X}^a_k \tag{7.26}$$

$$\Sigma^{yy}_{k|k} = \int [\gamma_k(\mathcal{X}_k) + v_k][\gamma_k(\mathcal{X}_k) + v_k]^T \mathcal{N}(\mathcal{X}^a_k; \hat{\mathcal{X}}^a_{k|k}, \Sigma^a_{k|k})d\mathcal{X}^a_k - \hat{\mathcal{Y}}_{k|k}\hat{\mathcal{Y}}^T_{k|k} \tag{7.27}$$

The predicted mean and error covariance of the delayed measurements are given by:

$$\hat{\mathcal{Z}}_{k+1|k} = (1-p)\hat{\mathcal{Y}}_{k+1|k} + p\hat{\mathcal{Y}}_{k|k}$$

and

$$\Sigma^{\mathcal{ZZ}}_{k+1|k} = (1-p)\Sigma^{yy}_{k+1|k} + p\Sigma^{yy}_{k|k} + p(1-p)(\hat{\mathcal{Y}}_{k+1|k} - \hat{\mathcal{Y}}_{k|k})(\hat{\mathcal{Y}}_{k+1|k} - \hat{\mathcal{Y}}_{k|k})^T.$$

The Kalman gain for noise is

$$K^v_{k+1} = \Sigma^{v\mathcal{Z}}_{k+1|k}(\Sigma^{\mathcal{ZZ}}_{k+1|k})^{-1},$$

where the cross error covariance is given by [176]

$$\Sigma^{v\mathcal{Z}}_{k+1|k} = (1-p)R_{k+1}.$$

Finally, the expressions for the posterior estimate of the measurement noise and its error covariance are

$$\hat{v}_{k+1|k+1} = K^v_{k+1}(\mathcal{Z}_{k+1} - \hat{\mathcal{Z}}_{k+1|k})$$

and

$$\Sigma^{vv}_{k+1|k+1} = R_{k+1} - K^v_{k+1}\Sigma^{\mathcal{ZZ}}_{k+1|k}(K^v_{k+1})^T.$$

7.3.3 State estimation

The prior estimate and its covariance can be written as

$$\hat{\mathcal{X}}_{k+1|k} = \int \phi_{k+1}(\mathcal{X}_k)\mathcal{N}(\mathcal{X}_k; \hat{\mathcal{X}}_{k|k}, \Sigma_{k|k})d\mathcal{X}_k \tag{7.28}$$

and

$$\Sigma_{k+1|k} = \int \phi_k(\mathcal{X}_k)\phi^T_k(\mathcal{X}_k)\mathcal{N}(\mathcal{X}_k; \hat{\mathcal{X}}_{k|k}, \Sigma_{k|k})d\mathcal{X}_k - \hat{\mathcal{X}}_{k+1|k}\hat{\mathcal{X}}^T_{k+1|k} + Q_k. \tag{7.29}$$

The cross error covariances are given by

$$\Sigma^{\mathcal{XY}}_{k+1,k|k} = \mathbb{E}[\tilde{\mathcal{X}}_{k+1|k}\tilde{\mathcal{Y}}^T_{k|k}|\mathcal{Z}_k]$$
$$= \int \phi_k(\mathcal{X}_k)[\gamma_k(\mathcal{X}_k) + v_k]^T\mathcal{N}(\mathcal{X}^a_k; \hat{\mathcal{X}}^a_{k|k}, \Sigma^a_{k|k})d\mathcal{X}^a_k - \hat{\mathcal{X}}_{k+1|k}\hat{\mathcal{Y}}^T_{k|k}$$
$$\tag{7.30}$$

and
$$\Sigma^{\mathcal{X}\mathcal{Y}}_{k+1|k} = \mathbb{E}[\tilde{\mathcal{X}}_{k+1|k}\tilde{\mathcal{Y}}^T_{k+1|k}|\mathcal{Z}_k]$$
$$= \int \mathcal{X}_{k+1}\gamma_{k+1}(\mathcal{X}_{k+1})^T \mathcal{N}(\mathcal{X}_{k+1}; \hat{\mathcal{X}}_{k+1|k}, \Sigma_{k+1|k})d\mathcal{X}_{k+1} - \hat{\mathcal{X}}_{k+1|k}\hat{\mathcal{Y}}^T_{k+1|k}.$$
$$(7.31)$$

The Kalman gain for state estimation is
$$K^{\mathcal{X}}_{k+1} = \Sigma^{\mathcal{X}\mathcal{Z}}_{k+1|k}(\Sigma^{\mathcal{Z}\mathcal{Z}}_{k+1|k})^{-1},$$
where
$$\Sigma^{\mathcal{X}\mathcal{Z}}_{k+1|k} = (1-\mathrm{p})\Sigma^{\mathcal{X}\mathcal{Y}}_{k+1|k} + p_{k+1}\Sigma^{\mathcal{X}\mathcal{Y}}_{k+1,k|k}.$$

Finally, the posterior state estimate and the posterior error covariance of the states are
$$\hat{\mathcal{X}}_{k+1|k+1} = \hat{\mathcal{X}}_{k+1|k} + K^{\mathcal{X}}_{k+1}(\mathcal{Z}_{k+1} - \hat{\mathcal{Z}}_{k+1|k})$$
and
$$\Sigma_{k+1|k+1} = \Sigma_{k+1|k} - K^{\mathcal{X}}_{k+1}\Sigma^{\mathcal{Z}\mathcal{Z}}_{k+1|k}(K^{\mathcal{X}}_{k+1})^T.$$

The cross covariance between the measurement noise and the state is
$$\Sigma^{\mathcal{X}v}_{k+1|k+1} = -K^{\mathcal{X}}_{k+1}\Sigma^{\mathcal{Z}\mathcal{Z}}_{k+1|k}(K^v_{k+1})^T.$$

The filtering with one-step randomly delayed measurements is performed by recursively performing the two steps discussed in Subsections 7.3.2 and 7.3.3. It should be noted that under the assumption that the measurements are not delayed i.e., $\beta_k = 0$, the delayed measurement equation reduces to the non-delayed measurement equation, and the formulated problem reduces to the ordinary state estimation problem.

The integrals which appear in Eqs. (7.24) to (7.27) and (7.28) to (7.31) are generally not available in closed-form and will be approximately evaluated. The accuracy of the filter depends on the accuracy of approximation of the integrals. Cubature and quadrature based rules are applied for evaluating the intractable integrals. See [143] for simulation examples.

7.4   Nonlinear filter for any arbitrary step randomly delayed measurement

Let us consider a nonlinear system given by a process and measurement equation, mentioned in the previous section. The expression for the $(N-1)$ step randomly delayed measurement at k^{th} time instant can be written as

$$\mathcal{Z}_k = (1-\beta_1)\mathcal{Y}_k + \beta_1(1-\beta_2)\mathcal{Y}_{k-1} + \beta_1\beta_2(1-\beta_3)\mathcal{Y}_{k-2}$$
$$+ \cdots + (\prod_{i=1}^{N-1}\beta_i)(1-\beta_N)\mathcal{Y}_{(k-N+1)},$$
$$(7.32)$$

where β_j $(j = 1, 2, \cdots, \mathbb{N})$ are the independent Bernoulli random variables taking values either 0 or 1 with the probability mentioned in Eq. (7.4). We write the above equation as

$$\mathcal{Z}_k = C_0 \mathcal{Y}_k + C_1 \mathcal{Y}_{k-1} + C_2 \mathcal{Y}_{k-2} + \cdots + C_{\mathbb{N}-1} \mathcal{Y}_{(k-\mathbb{N}+1)}, \tag{7.33}$$

where

$$C_i = (\prod_{j=0}^{i} \beta_j)(1 - \beta_{i+1}), \tag{7.34}$$

with $\beta_0 = 1$. Hence,

$$\begin{aligned} \mathcal{Z}_k &= \sum_{i=0}^{\mathbb{N}-1} C_i \mathcal{Y}_{k-i} \\ &= \sum_{i=0}^{\mathbb{N}-1} (\prod_{j=0}^{i} \beta_j)(1 - \beta_{i+1}) \mathcal{Y}_{k-i}. \end{aligned} \tag{7.35}$$

The values of C_i will be either 0 or 1. At any specific instant, at most one C_i can be 1 while all the remaining will be 0. If $C_j = 1$ and $C_i = 0$ $(\forall \ i \neq j)$, then the received measurement will be delayed by j time steps. If all the C_i $(\forall \ 0 \leq i \leq \mathbb{N} - 1)$ are zero, it represents that no measurement is received at the corresponding time instant. We exclude such a possibility here. Even if that happens, the measurement update step can be skipped and only the process update step can be executed.

Lemma 4 *The probability that the measurement received at the kth time step is actually i time step delayed, can be given as*

$$\alpha_i = p^i(1 - p). \tag{7.36}$$

Proof:
α_i can be given as

$$\alpha_i = \mathbb{E}[C_i] = \mathbb{E}[(\prod_{j=0}^{i} \beta_j)(1 - \beta_{i+1})].$$

As β_j is independent of β_k $\forall \ j \neq k$, hence

$$\mathbb{E}[(\prod_{j=0}^{i} \beta_j)(1 - \beta_{i+1})] = \mathbb{E}[(\prod_{j=0}^{i} \beta_j)] \mathbb{E}[(1 - \beta_{i+1})]. \tag{7.37}$$

Also,

$$\mathbb{E}[(\prod_{j=0}^{i} \beta_j)] = (\prod_{j=0}^{i} \mathbb{E}[\beta_j]) = p^i, \tag{7.38}$$

and
$$\mathbb{E}[(1 - \beta_{i+1})] = (1 - \mathbf{p}). \tag{7.39}$$

Substituting (7.38) and (7.39) into (7.37) , we get

$$\mathbb{E}[(\prod_{j=0}^{i} \beta_j)(1 - \beta_{i+1})] = \mathbf{p}^i(1 - \mathbf{p}). \tag{7.40}$$

Hence,
$$\alpha_i = \mathbb{E}[C_i] = \mathbf{p}^i(1 - \mathbf{p}). \tag{7.41}$$

∎

Lemma 5 *The probability that no measurement will be received at k^{th} time instant can be given as $\tilde{P}_k = \mathbf{p}^N$.*

Proof: $\tilde{P}_k$ can be given as

$$\tilde{P}_k = \mathbb{E}[C] = \mathbb{E}[1 - \sum_{i=0}^{N-1} C_i] = 1 - \sum_{i=0}^{N-1} \mathbb{E}[C_i].$$

Replacing the value of $\mathbb{E}[C_i]$ from equation (7.41), we get

$$\tilde{P}_k = 1 - \sum_{i=0}^{N-1} \mathbf{p}^i(1 - \mathbf{p}) = \mathbf{p}^N.$$

∎

It should be noted that as $\mathbf{p}$ is a fraction, the probability that no measurement arrives at k^{th} time instant i.e., $\tilde{P}_k = \mathbf{p}^N$ will be very small, if N is chosen sufficiently large. It justifies our assumption of neglecting missing measurements.

Our objective is to estimate the states of a system, $\hat{\mathcal{X}}$ from the delayed measurement $\mathcal{Z}_k$. As the system is nonlinear, the posterior and prior pdfs will be non-Gaussian. However, here we approximate them as Gaussian and represented with mean and covariance. With the help of the non-delayed Gaussian filter, discussed in earlier chapters, the $\hat{\mathcal{Y}}_{k|k-1}$, $\Sigma_{k|k-1}^{xy}$, and $\Sigma_{k|k-1}^{yy}$ can be calculated. Further, the following Lemmas provide the formula to calculate $\hat{\mathcal{Z}}_{k|k-1}$, $\Sigma_{k|k-1}^{xz}$, and $\Sigma_{k|k-1}^{zz}$ which finally lead to $\hat{\mathcal{X}}_{k|k}$ [144].

Lemma 6 *If no measurement is lost for $(N-1)$ delay, the estimate of measurement at k^{th} time instant can be given as*

$$\hat{\mathcal{Z}}_{k|k-1} = (1 - p) \sum_{i=0}^{N-1} p^i \hat{\mathcal{Y}}_{k-i|k-1}. \tag{7.42}$$

Proof:
From Eq. (7.35)

$$
\begin{aligned}
\hat{\mathcal{Z}}_{k|k-1} &= \mathbb{E}[\mathcal{Z}_k] \\
&= \mathbb{E}\left[\sum_{i=0}^{N-1}(\prod_{j=0}^{i}\beta_j)(1-\beta_{i+1})\mathcal{Y}_{k-i}\right] \\
&= \sum_{i=0}^{N-1}\mathbb{E}[(\prod_{j=0}^{i}\beta_j)(1-\beta_{i+1})]\hat{\mathcal{Y}}_{k-i|k-1}.
\end{aligned}
\tag{7.43}
$$

Substituting (7.40) into (7.43), we get

$$
\hat{\mathcal{Z}}_{k|k-1} = (1-p)\sum_{i=0}^{N-1}p^i\hat{\mathcal{Y}}_{k-i|k-1}.
$$

■

Lemma 7 *If no measurement is lost for* $(N-1)$ *delay, the covariance of measurement at* k^{th} *time instant can be given as*

$$
\begin{aligned}
\Sigma^{\mathcal{ZZ}}_{k|k-1} &= (1-p)\sum_{i=0}^{N-1}p^i\Sigma^{\mathcal{YY}}_{k-i|k-1} \\
&+ (1-p)\sum_{i=0}^{N-1}p^i\left(1-p^i(1-p)\right)((\hat{\mathcal{Y}}_{k-i|k-1})(\hat{\mathcal{Y}}_{k-i|k-1})^T).
\end{aligned}
\tag{7.44}
$$

Proof:
$$
\Sigma^{\mathcal{ZZ}}_{k|k-1} = \mathbb{E}[(\mathcal{Z}_k - \hat{\mathcal{Z}}_{k|k-1})(\mathcal{Z}_k - \hat{\mathcal{Z}}_{k|k-1})^T]
$$
From equations (7.35) and (7.42), we get

$$
\begin{aligned}
(\mathcal{Z}_k - \hat{\mathcal{Z}}_{k|k-1}) &= \sum_{i=0}^{N-1}(\prod_{j=0}^{i}\beta_j)(1-\beta_{i+1})\mathcal{Y}_{k-i} - (1-p)\sum_{i=0}^{N-1}p^i\hat{\mathcal{Y}}_{k-i|k-1} \\
&= \underbrace{\sum_{i=0}^{N-1}(\prod_{j=0}^{i}\beta_j)(1-\beta_{i+1})(\mathcal{Y}_{k-i} - \hat{\mathcal{Y}}_{k-i|k-1})}_{M_1} \\
&+ \underbrace{\sum_{i=0}^{N-1}(\prod_{j=0}^{i}\beta_j(1-\beta_{i+1}) - (1-p)p^i)\hat{\mathcal{Y}}_{k-i|k-1}}_{M_2}.
\end{aligned}
$$

Hence,

$$
\Sigma^{\mathcal{ZZ}}_{k|k-1} = \mathbb{E}[M_1M_1^T] + \mathbb{E}[M_1M_2^T] + \mathbb{E}[M_2M_1^T] + \mathbb{E}[M_2M_2^T].
\tag{7.45}
$$

Now,

$$\mathbb{E}[M_1 M_1^T] = \sum_{i=0}^{N-1} \mathbb{E}[(\prod_{j=0}^{i} \beta_j)(1-\beta_{i+1})]^2 \mathbb{E}[(\mathcal{Y}_{k-i} - \hat{\mathcal{Y}}_{k-i|k-1})(\mathcal{Y}_{k-i} - \hat{\mathcal{Y}}_{k-i|k-1})^T]$$

$$= \sum_{i=0}^{N-1} \mathbb{E}[(\prod_{j=0}^{i} \beta_j)(1-\beta_{i+1})] \times \Sigma_{k-i|k-1}^{\mathcal{Y}\mathcal{Y}}.$$

Substituting equations (7.38) and (7.39) into (7.37) and then using the result with above equation, we get

$$\mathbb{E}[M_1 M_1^T] = (1-\mathbf{p}) \sum_{i=0}^{N-1} \mathbf{p}^i \Sigma_{k-i|k-1}^{\mathcal{Y}\mathcal{Y}}. \tag{7.46}$$

Next,

$$\mathbb{E}[M_1 M_2^T] = \sum_{s=0}^{N-1} \sum_{t=0}^{N-1} \mathbb{E}[A_s A_t^T], \tag{7.47}$$

where

$$A_s = (\prod_{j=0}^{s} \beta_j)(1-\beta_{s+1})(\mathcal{Y}_{k-s} - \hat{\mathcal{Y}}_{k-s|k-1})$$

and

$$A_t = (\prod_{j=0}^{t} \beta_j(1-\beta_{t+1}) - (1-\mathbf{p})\mathbf{p}^t)\hat{\mathcal{Y}}_{k-t|k-1}.$$

For any s and t,

$$\mathbb{E}[A_s A_t^T]$$

$$= \mathbb{E}[(\prod_{j=0}^{s} \beta_j)(1-\beta_{s+1})(\prod_{j=0}^{t} \beta_j \cdots (1-\beta_{t+1}) - (1-\mathbf{p})\mathbf{p}^t)((\mathcal{Y}_{k-s})(\hat{\mathcal{Y}}_{k-t|k-t-1})^T)]$$

$$- \mathbb{E}[(\prod_{j=0}^{s} \beta_j)(1-\beta_{s+1})(\prod_{j=0}^{t} \beta_j(1-\beta_{t+1}) \cdots - (1-\mathbf{p})\mathbf{p}^t)((\hat{\mathcal{Y}}_{k-s|k-s-1})(\hat{\mathcal{Y}}_{k-t|k-1})^T)]$$

$$\tag{7.48}$$

or,

$$\mathbb{E}[A_s A_t^T]$$

$$= \mathbb{E}[(\prod_{j=0}^{s} \beta_j)(1-\beta_{s+1})(\prod_{j=0}^{t} \beta_j \cdots (1-\beta_{t+1}) - (1-\mathbf{p})\mathbf{p}^t) - (\prod_{j=0}^{s} \beta_j)$$

$$(1-\beta_{s+1})(\prod_{j=0}^{t} \beta_j(1-\beta_{t+1}) - (1-\mathbf{p})\mathbf{p}^t)]((\hat{\mathcal{Y}}_{k-s|k-1})(\hat{\mathcal{Y}}_{k-t|k-1})^T) \tag{7.49}$$

or,

$$\mathbb{E}[A_s A_t^T] = 0. \tag{7.50}$$

Substituting (7.50) into (7.47),

$$\mathbb{E}[M_1 M_2^T] = 0. \tag{7.51}$$

Similarly, it can be shown that

$$\mathbb{E}[M_2 M_1^T] = 0. \tag{7.52}$$

Now,

$$\mathbb{E}[M_2 M_2^T]$$

$$= \sum_{i=0}^{N-1} \mathbb{E}[(\prod_{j=0}^{i} \beta_j (1 - \beta_{i+1}) - (1 - \mathrm{p})\mathrm{p}^i)^2] (\hat{\mathcal{Y}}_{k-i|k-1})(\hat{\mathcal{Y}}_{k-i|k-1})^T$$

$$= \sum_{i=0}^{N-1} \mathbb{E}[(\prod_{j=0}^{i} \beta_j (1 - \beta_{i+1}))^2 + ((1 - \mathrm{p})\mathrm{p}^i)^2 - 2(\prod_{j=0}^{i} \beta_j (1 - \beta_{i+1}))((1 - \mathrm{p})\mathrm{p}^i)]$$

$$\times (\hat{\mathcal{Y}}_{k-i|k-1})(\hat{\mathcal{Y}}_{k-i|k-1})^T. \tag{7.53}$$

In the above expression

$$\mathbb{E}[(\prod_{j=0}^{i} \beta_j (1 - \beta_{i+1}))^2] = \mathbb{E}[(\prod_{j=0}^{i} \beta_j (1 - \beta_{i+1}))].$$

Substituting Eq. (7.38) and (7.39) into (7.37) and then the result into the above equation, we get

$$\mathbb{E}[(\prod_{j=0}^{i} \beta_j (1 - \beta_{i+1}))^2] = \mathrm{p}^i (1 - \mathrm{p}). \tag{7.54}$$

Hence,

$$\mathbb{E}[M_2 M_2^T] = \sum_{i=0}^{N-1} \left(\mathrm{p}^i (1 - \mathrm{p}) - \mathrm{p}^{2i} (1 - \mathrm{p})^2 \right) \left((\hat{\mathcal{Y}}_{k-i|k-1})(\hat{\mathcal{Y}}_{k-i|k-1})^T \right). \tag{7.55}$$

Substituting (7.46), (7.51), (7.52) and (7.55) into (7.45), we obtain

$$\Sigma_{k|k-1}^{\mathcal{Z}\mathcal{Z}} = (1 - \mathrm{p}) \sum_{i=0}^{N-1} \mathrm{p}^i \Sigma_{k-i|k-1}^{\mathcal{Y}\mathcal{Y}} + \sum_{i=0}^{N-1} \left(\mathrm{p}^i (1 - \mathrm{p}) - \mathrm{p}^{2i} (1 - \mathrm{p})^2 \right)$$

$$\times ((\hat{\mathcal{Y}}_{k-i|k-1})(\hat{\mathcal{Y}}_{k-i|k-1})^T)$$

or,

$$\Sigma_{k|k-1}^{\mathcal{Z}\mathcal{Z}} = (1-\mathrm{p}) \sum_{i=0}^{N-1} \mathrm{p}^i \Sigma_{k-i|k-1}^{\mathcal{Y}\mathcal{Y}} + (1-\mathrm{p}) \sum_{i=0}^{N-1} \mathrm{p}^i \left(1 - \mathrm{p}^i(1-\mathrm{p})\right)$$
$$\times ((\hat{\mathcal{Y}}_{k-i|k-1})(\hat{\mathcal{Y}}_{k-i|k-1})^T).$$

■

Lemma 8 *If no measurement is lost for (N − 1)-delay, the cross-covariance between the state and measurement at k^{th} time instant can be given as*

$$\Sigma_{k|k-1}^{\mathcal{X}\mathcal{Z}} = (1-p) \sum_{i=0}^{N-1} p^i \Sigma_{k-i|k-1}^{\mathcal{X}\mathcal{Y}}. \tag{7.56}$$

Proof:

$$\begin{aligned}
\Sigma_{k|k-1}^{\mathcal{X}\mathcal{Z}} &= \mathbb{E}[(\mathcal{X}_k - \hat{\mathcal{X}}_{k|k-1})(\mathcal{Z}_k - \hat{\mathcal{Z}}_{k|k-1})^T] \\
&= \mathbb{E}[(\mathcal{X}_k - \hat{\mathcal{X}}_{k|k-1})(M_1 + M_2)^T] \\
&= \mathbb{E}[(\mathcal{X}_k - \hat{\mathcal{X}}_{k|k-1})(M_1)^T] + \mathbb{E}[(\mathcal{X}_k - \hat{\mathcal{X}}_{k|k-1})(M_2)^T].
\end{aligned} \tag{7.57}$$

Now, replacing the value of M_1, we get

$$\mathbb{E}[(\mathcal{X}_k - \hat{\mathcal{X}}_{k|k-1})(M_1)^T]$$
$$= \mathbb{E}[(\mathcal{X}_k - \hat{\mathcal{X}}_{k|k-1}) \times (\sum_{i=0}^{N-1} (\prod_{j=0}^{i} \beta_j)(1 - \beta_{i+1})(\mathcal{Y}_{k-i} - \hat{\mathcal{Y}}_{k-i|k-1}))^T] \tag{7.58}$$
$$= (1-\mathrm{p}) \sum_{i=0}^{N-1} \mathrm{p}^i \mathbb{E}[(\mathcal{X}_k - \hat{\mathcal{X}}_{k|k-1})(\mathcal{Y}_{k-i} - \hat{\mathcal{Y}}_{k-i|k-1})^T]$$

or,

$$\mathbb{E}[(\mathcal{X}_k - \hat{\mathcal{X}}_{k|k-1})(M_1)^T] = (1-\mathrm{p}) \sum_{i=0}^{N-1} \mathrm{p}^i \Sigma_{k-i|k-1}^{\mathcal{X}\mathcal{Y}}. \tag{7.59}$$

Similarly replacing M_2, we get

$$\mathbb{E}[(\mathcal{X}_k - \hat{\mathcal{X}}_{k|k-1})(M_2)^T]$$
$$= \mathbb{E}[(\mathcal{X}_k - \hat{\mathcal{X}}_{k|k-1})(\sum_{i=0}^{N-1} (\prod_{j=0}^{i} \beta_j(1 - \beta_{i+1}) - (1 - \mathrm{p})\mathrm{p}^i)\hat{\mathcal{Y}}_{k-i|k-1})^T]$$
$$= \sum_{i=0}^{N-1} \mathbb{E}[(\prod_{j=0}^{i} \beta_j(1 - \beta_{i+1}) - (1 - \mathrm{p})\mathrm{p}^i)((\mathcal{X}_k - \hat{\mathcal{X}}_{k|k-1})\hat{\mathcal{Y}}_{k-i|k-1})^T] \tag{7.60}$$
$$= \sum_{i=0}^{N-1} \mathbb{E}[((1 - \mathrm{p})\mathrm{p}^i - (1 - \mathrm{p})\mathrm{p}^i)((\mathcal{X}_k - \hat{\mathcal{X}}_{k|k-1})\hat{\mathcal{Y}}_{k-i|k-1})^T]$$
$$= 0$$

Substituting (7.59) and (7.60) into (7.57), we get

$$\Sigma^{\mathcal{X}\mathcal{Z}}_{k|k-1} = (1 - \mathrm{p}) \sum_{i=0}^{N-1} \mathrm{p}^i \Sigma^{\mathcal{X}\mathcal{Y}}_{k-i|k-1}.$$

■

It is easy to verify that under the condition of no delay $N-1 = 0$, $\mathrm{p} = 0$ and $\mathcal{Z}_k = \mathcal{Y}_k$. In such a case the reader can easily verify that $\hat{\mathcal{Z}}_{k+1|k} = \hat{\mathcal{Y}}_{k+1|k}$, $\Sigma^{\mathcal{Z}\mathcal{Z}}_{k+1|k} = \Sigma^{\mathcal{Y}\mathcal{Y}}_{k+1|k}$, $\Sigma^{\mathcal{X}\mathcal{Z}}_{k+1|k} = \Sigma^{\mathcal{X}\mathcal{Y}}_{k+1|k}$ and the new formulation merges with the ordinary nonlinear filter. In the next subsection, we provide a detailed algorithm of the delay filter.

7.4.1   Algorithm

The algorithm for the deterministic sample point filter for arbitrary randomly delayed measurements is as follows.

STEP 1: Filter initialization

- Filter is initialized with $\hat{\mathcal{X}}_{0|0}$, $\Sigma_{0|0}$.

STEP 2: Prediction Step

- Represent $\mathcal{N}(\mathcal{X}_{k|k}; \hat{\mathcal{X}}_{k|k}, \Sigma_{k|k})$ with deterministic sample points, $\xi_{j,k|k}$ and their corresponding weights w_j (UKF/CQKF/GHF etc., method)

- Time update the sample points:

$$\xi_{j,k+1|k} = \phi_k(\xi_{j,k|k}).$$

- Compute the prior mean and error covariance:

$$\hat{\mathcal{X}}_{k+1|k} = \sum_{j=1}^{n_s} w_j \xi_{j,k+1|k},$$

$$\Sigma_{k+1|k} = \sum_{j=1}^{n_s} w_j (\xi_{j,k+1|k} - \hat{\mathcal{X}}_{k+1|k})(\xi_{j,k+1|k} - \hat{\mathcal{X}}_{k+1|k})^T + Q_k.$$

STEP 3: Measurement Update

- Represent $\mathcal{N}(\mathcal{X}_{k+1|k}; \hat{\mathcal{X}}_{k+1|k}, \Sigma_{k+1|k})$ with sample points, $\xi_{j,k+1|k}$ and their corresponding weights w_j.

- Compute the predicted measurement without delay

$$\hat{\mathcal{Y}}_{k+1|k} = \sum_{j=1}^{n_s} w_j \gamma_k(\xi_{j,k+1|k}).$$

- Compute the expected delayed measurement:

$$\hat{\mathcal{Z}}_{k+1|k} = (1 - \mathrm{p}) \sum_{i=0}^{N-1} \mathrm{p}^i \hat{\mathcal{Y}}_{k+1-i|k}.$$

- Compute the measurement error covariance:

$$\Sigma_{k+1|k}^{\mathcal{Y}\mathcal{Y}} = \sum_{j=1}^{n_s} w_j (\gamma_k(\xi_{j,k+1|k}) - \hat{\mathcal{Y}}_{k+1|k})(\gamma_k(\xi_{j,k+1|k}) - \hat{\mathcal{Y}}_{k+1|k})^T + R_k.$$

- Calculate $\Sigma_{k+1|k}^{\mathcal{Z}\mathcal{Z}}$ with the help of (7.44).

- Calculate the cross covariances:

$$\Sigma_{k+1|k}^{\mathcal{X}\mathcal{Y}} = \sum_{j=1}^{n_s} w_j (\xi_{j,k+1|k} - \hat{\mathcal{X}}_{k+1|k})(\gamma_k(\xi_{j,k+1|k}) - \hat{\mathcal{Y}}_{k+1|k})^T,$$

$$\Sigma_{k+1|k}^{\mathcal{X}\mathcal{Z}} = (1 - \mathrm{p}) \sum_{i=0}^{N-1} \mathrm{p}^i \Sigma_{k+1-i|k}^{\mathcal{X}\mathcal{Y}}.$$

- Calculate the Kalman gain:

$$K_{k+1} = \Sigma_{k+1|k}^{\mathcal{X}\mathcal{Z}} (\Sigma_{k+1|k}^{\mathcal{Z}\mathcal{Z}})^{-1}.$$

- Compute the posterior estimate:

$$\hat{\mathcal{X}}_{k+1|k+1} = \hat{\mathcal{X}}_{k+1|k} + K_{k+1}(\mathcal{Z}_{k+1} - \hat{\mathcal{Z}}_{k+1|k}).$$

- Compute the posterior error covariance:

$$\Sigma_{k+1|k+1} = \Sigma_{k+1|k} - K_{k+1}\Sigma_{k+1|k}^{\mathcal{Z}\mathcal{Z}}K_{k+1}^T.$$

7.5   Simulation

Let us recall the estimation problem of multiple superimposed sinusoids, discussed in Chapter 4. The problem has been solved here with the consideration that random delay in measurement occurs. Here, the true values are generated using the initial estimate $\mathcal{X}_0 = [200\ 1000\ 2000\ 2\ 2\ 2]^T$ and a sampling time period $T = 0.25msec$. For filtering purposes, the initial estimate and error covariance are considered as $[216\ 1080\ 2160\ 1.8\ 3\ 2.4]^T$ and $diag([20^2\ 20^2$

20^2 0.5 0.5 0.5]) respectively. The process noise covariance is considered as $Q = diag([\sigma_f^2 \; \sigma_f^2 \; \sigma_f^2 \; \sigma_a^2 \; \sigma_a^2 \; \sigma_a^2])$ where $\sigma_f^2 = 5.1\mu Hz^2/ms^2$ and $\sigma_a^2 = 8\mu V^2/ms^2$. The measurement noise covariance is taken as $R = diag([0.009Volt^2 \quad 0.009Volt^2])$.

The states are estimated for 500 steps with the arbitrary step random delay estimator where the quadrature points are generated with the CKF algorithm [98] and the results are compared in terms of the RMSEs (mentioned in Eq. (5.38)) computed over 500 Monte Carlo runs . Figures 7.1(a) and 7.1(b) represent the RMSE plots of frequency and amplitude respectively, while Figures 7.1(c) and 7.1(d) show the average RMSE plotted against the probability, p, for the case of one delay in measurement. Similarly, Figures 7.2(a) to 7.2(d) represent the same plots for the two delay case. The RMSE plots show that, in the presence of delay, the filter designed to account for it provides more accurate results than a filter which is designed for delay-free measurements.

7.6   Summary

This chapter studies the state estimation problem in presence of randomly delayed measurements. Here, a random delay in the measurement is modelled with a Bernoulli random variable. In such a scenario, a closed form solution is derived for a linear system with one step randomly delayed measurements. For a nonlinear system and with one step randomly delayed measurement, quadrature filters are modified to carry out the state estimation efficiently. Further, a more generalized nonlinear system has been considered where the measurements are delayed randomly by an arbitrary number of time steps and an approximate solution to filtering problems in such systems using quadrature filters is proposed. The algorithm presented is demonstrated using a simulation example.

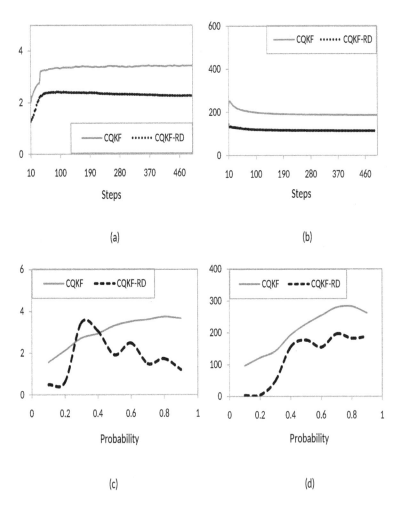

(a)   (b)

(c)   (d)

FIGURE 7.1: One delay with latency probability **p** = 0.4: (a) RMSE of amplitude in V (b) RMSE of frequency in Hz (c) averaged RMSE vs. probability plot for amplitude (d) averaged RMSE vs. probability plot for frequency [144].

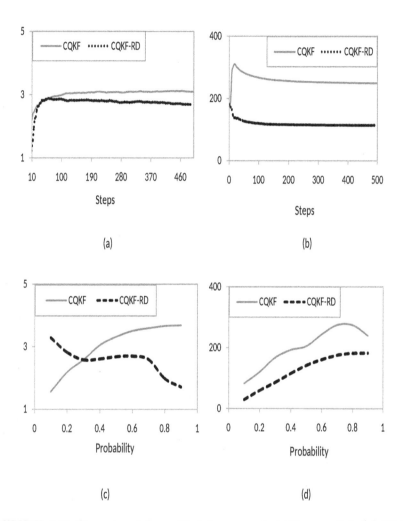

FIGURE 7.2: Two step delay with latency probability $p = 0.4$: (a) RMSE of amplitude in V (b) RMSE of frequency in Hz (c) averaged RMSE vs. probability plot for amplitude (d) averaged RMSE vs. probability plot for frequency [144].

Chapter 8

Continuous-discrete filtering

8.1	Introduction	159
8.2	Continuous time filtering	160
	8.2.1 Continuous filter for a linear Gaussian system	161
	8.2.2 Nonlinear continuous time system	167
	8.2.2.1 The extended Kalman-Bucy filter	167
8.3	Continuous-discrete filtering	168
	8.3.1 Nonlinear continuous time process model	171
	8.3.2 Discretization of process model using Runge-Kutta method	172
	8.3.3 Discretization using Ito-Taylor expansion of order 1.5 ...	172
	8.3.4 Continuous-discrete filter with deterministic sample points	174
8.4	Simulation examples	176
	8.4.1 Single dimensional filtering problem	176
	8.4.2 Estimation of harmonics	177
	8.4.3 RADAR target tracking problem	178
8.5	Summary	186

8.1 Introduction

Until now we have considered the state estimation problem for a discrete system. A system is discrete if both the process and the measurement equations are expressed in a discrete time domain. Due to advances in computer hardware, and the recursive nature of the solution, discrete filtering is very popular. However, in many real life problems, process models are continuous because they are derived from the law of physics and a measurement equation is discrete because sensor outputs are sampled. Such a state space system where state dynamics are modeled as continuous time stochastic processes, and the continuous time states are observed discretely in measurement model, is known as a continuous-discrete system.

The main objective of this chapter is to present various existing techniques of state estimation for a nonlinear continuous-discrete system. The general approach to solve such a problem is to discretize the process model using the Ito-Taylor expansion or some other means and incorporating it in the framework of a discrete filter. The nonlinear filtering problems of this kind commonly appear in several real-life problems like target tracking [13], navigation [59], telecommunication [171], etc.

8.2 Continuous time filtering

In many physical systems, the process is described by a continuous time stochastic differential equation and measurements are also received continuously. For example, in analog communication systems [171], analog receivers are the measuring devices which demodulate the transmitted continuous time signal. Here, the measured signals are continuous time processes. Further, in many analog control systems which operate without digital computers, the measured signals are in the continuous time domain [153]. Recursive estimation of this kind of system is done by continuous time filtering. In general, a continuous time process and measurement equation can be represented with an Ito or Stratonovich type stochastic differential equation given by

$$d\mathcal{X}(t) = f(\mathcal{X}(t), \mathcal{U}(t), t)dt + L(t)d\beta(t) \tag{8.1}$$

and

$$dy(t) = g(\mathcal{X}(t), t)dt + V(t)d\zeta(t), \tag{8.2}$$

where $\mathcal{X}(t) \in \mathbb{R}^n$ is the state of the system, $y(t) \in \mathbb{R}^m$ is the measured value, $f(\cdot)$ is the drift function, $g(\cdot)$ is the measurement function, $L(t)$ and $V(t)$ are time varying matrices, $\beta(t)$ and $\zeta(t)$ are mutually independent Brownian motions with diagonal diffusion matrices $Q(t)$ and $R(t)$ respectively.

The above equations can further be written as

$$\dot{\mathcal{X}}(t) = \frac{d\mathcal{X}(t)}{dt} = f(\mathcal{X}(t), \mathcal{U}(t), t) + L(t)\eta(t), \tag{8.3}$$

and

$$\mathcal{Y}(t) = \frac{dy(t)}{dt} = g(\mathcal{X}(t), t) + v(t), \tag{8.4}$$

where $\eta(t) = d\beta(t)/dt$, $v(t) = d\zeta/dt$ are white noises (i.e., formal derivatives of the Brownian motion) and $\mathcal{Y}$ is the differential measurement. Without loss of generality, we assume $V(t) = I$. The white noises $\eta(t)$ and $v(t)$ are zero mean and covariance $Q_c(t)$ and $R_c(t)$ respectively. Note that Brownian motion has non-differentiable sample paths, although the above notational convention, which transforms the diffusion process into a familiar 'ODE with noise' form, is still commonly used in engineering. The purpose of the continuous time filtering is to compute, if possible recursively, the posterior distribution of states i.e., $p(\mathcal{X}(t)|\{\mathcal{Y}(\tau) : 0 \leq \tau \leq t\})$. Often it is not possible to find out the expression for the probability density function. In such cases, expressions for relevant moments are calculated.

8.2.1 Continuous filter for a linear Gaussian system

For a linear system and white Gaussian noises, a closed form solution can be derived which is popularly known as the Kalman-Bucy filter [90]. A continuous time linear system can be expressed as

$$\dot{\mathcal{X}}(t) = A(t)\mathcal{X}(t) + B(t)\mathcal{U}(t) + L(t)\eta(t) \tag{8.5}$$

and

$$\mathcal{Y}(t) = C(t)\mathcal{X}(t) + v(t), \tag{8.6}$$

where $B(t)$ is the input matrix and $\mathcal{U}(t)$ is the input to the system. The filtering problem can be formulated as finding the estimate of states $\hat{\mathcal{X}}(t)$ which minimizes the mean square error $\mathbb{E}[(\mathcal{X}(t) - \hat{\mathcal{X}}(t))(\mathcal{X}(t) - \hat{\mathcal{X}}(t))^T]$ given $\{\mathcal{Y}(\tau) : 0 \leq \tau \leq t\}$. The recursive estimation algorithm of the Kalman-Bucy filter is summarized in the following Theorem [113].

Theorem 4 *For a system described by Eqs. (8.5) and (8.6), the optimal estimator has the form of*

$$\dot{\hat{\mathcal{X}}}(t) = A(t)\hat{\mathcal{X}}(t) + B(t)\mathcal{U}(t) + K(t)(\mathcal{Y}(t) - C(t)\hat{\mathcal{X}}(t)), \tag{8.7}$$

where $K(t)$ is the Kalman gain whose expression is given by $K(t) = \Sigma(t)C(t)^T R_c^{-1}$. $\Sigma(t)$ is the error covariance matrix i.e., $\Sigma(t) = \mathbb{E}[(\mathcal{X}(t) - \hat{\mathcal{X}}(t))(\mathcal{X}(t) - \hat{\mathcal{X}}(t))^T]$ which satisfies

$$\dot{\Sigma}(t) = A(t)\Sigma(t) + \Sigma(t)A(t)^T - \Sigma(t)C(t)^T R_c(t)^{-1} C(t)\Sigma(t) + L(t)Q_c(t)L(t)^T. \tag{8.8}$$

Proof 4 *We assume the estimator is in the form of (8.7). Subtracting Eq. (8.7) from (8.5), we get*

$$\dot{\mathcal{X}}(t) - \dot{\hat{\mathcal{X}}}(t) = (A(t) - K(t)C(t))(\mathcal{X}(t) - \hat{\mathcal{X}}(t)) + L(t)\eta(t) - K(t)v(t)$$

or,

$$\dot{e}(t) = (A(t) - K(t)C(t))e(t) + L(t)\eta(t) - K(t)v(t). \tag{8.9}$$

Now,

$$\dot{e}(t) = (A(t) - K(t)C(t))e(t) + \xi(t), \tag{8.10}$$

where $\xi(t) = L(t)\eta(t) - K(t)v(t)$. Now the expression for error covariance becomes

$$\Sigma(t) = \mathbb{E}[e(t)e(t)^T].$$

or,

$$\begin{aligned}
\dot{\Sigma}(t) &= \mathbb{E}[\dot{e}(t)e(t)^T + e(t)\dot{e}(t)^T] \\
&= \mathbb{E}[(A(t) - K(t)C(t))e(t)e(t)^T + \xi(t)e(t)^T \\
&\quad + e(t)e(t)^T(A(t) - K(t)C(t))^T + e(t)\xi(t)^T] \\
&= (A(t) - K(t)C(t))\Sigma(t) + \Sigma(t)(A(t) - K(t)C(t))^T \\
&\quad + \mathbb{E}[\xi(t)e(t)^T + e(t)\xi(t)^T]
\end{aligned} \tag{8.11}$$

The error dynamic Eq. (8.10) can be solved [10, 118]

$$e(t) = \phi(t, t_0)e(t_0) + \int_{t_0}^{t} \phi(t, \tau)\xi(\tau)d\tau, \tag{8.12}$$

where $\phi(\cdot)$ is the state transition matrix of $(A(t) - K(t)C(t))$. The term $\mathbb{E}[e(t)\xi(t)^T]$ becomes

$$\mathbb{E}[e(t)\xi(t)^T] = \phi(t, t_0)\mathbb{E}[e(t_0)\xi(t)^T] + \int_{t_0}^{t} \phi(t, \tau)\mathbb{E}[\xi(\tau)\xi(t)^T]d\tau$$

$$= \int_{t_0}^{t} \phi(t, \tau)\mathbb{E}[\xi(\tau)\xi(\tau)^T \delta(t - \tau)]d\tau \quad (as\ \mathbb{E}[e(t_0)\xi(t)^T] = 0). \tag{8.13}$$

We can write

$$\mathbb{E}[\xi(\tau)\xi(\tau)^T] = \mathbb{E}[(L(\tau)\eta(\tau) - K(\tau)v(\tau))(L(\tau)\eta(\tau) - K(\tau)v(\tau))^T]$$

$$= R_\xi(\tau) = L(\tau)Q_c(\tau)L^T(\tau) + K(\tau)R_c(\tau)K^T(\tau). \tag{8.14}$$

Substituting the above equation into (8.13), and using the relation $\int_a^b f(x)\delta(b - x)dx = 1/2f(b)$ [10] (Eq. 81), and the property of the state transition matrix, $\phi(t, t) = I$ [118] we obtain,

$$\mathbb{E}[e(t)\xi(t)^T] = \int_{t_0}^{t} \phi(t, \tau)R_\xi(\tau)\delta(t - \tau)]d\tau$$

$$= 1/2R_\xi(t)\phi(t, t) \tag{8.15}$$

$$= 1/2R_\xi(t)$$

Similarly, it can be proved that $\mathbb{E}[\xi(t)e(t)^T] = 1/2R_\xi(t)$. Substituting those values, (8.11) becomes

$$\dot{\Sigma}(t) = (A(t) - K(t)C(t))\Sigma(t) + \Sigma(t)(A(t) - K(t)C(t))^T + R_\xi(t)$$

$$= (A(t) - K(t)C(t))\Sigma(t) + \Sigma(t)(A(t) - K(t)C(t))^T \tag{8.16}$$

$$+ L(t)Q_c(t)L^T(t) + K(t)R_c(t)K^T(t)$$

The above equation can further be simplified

$$\dot{\Sigma}(t) = A(t)\Sigma(t) + \Sigma(t)A(t)^T + L(t)Q_c(t)L^T(t)$$

$$- K(t)C(t)\Sigma(t) - \Sigma(t)C(t)^T K(t)^T + K(t)R_c(t)K(t)^T$$

$$+ \Sigma(t)C(t)^T R_c(t)^{-1}C(t)\Sigma(t) - \Sigma(t)C(t)^T R_c(t)^{-1}C(t)\Sigma(t)$$

$$= A(t)\Sigma(t) + \Sigma(t)A(t)^T + L(t)Q_c(t)L^T(t) + (K(t)R_c(t) - \Sigma(t)C(t)^T) \times$$

$$R_c(t)^{-1}(K(t)R_c(t) - \Sigma(t)C(t)^T)^T - \Sigma(t)C(t)^T R_c(t)^{-1}C(t)\Sigma(t) \tag{8.17}$$

We need to find $K(t)$ such that $\Sigma(t)$ is as small as possible, which can be done by choosing $K(t)$ so that $\dot{\Sigma}(t)$ decreases by the maximum amount possible at each instant of time. This can be accomplished by setting

$$K(t)R_c(t) - \Sigma(t)C(t)^T = 0,$$

or,

$$K(t) = \Sigma(t)C(t)^T R_c(t)^{-1}. \tag{8.18}$$

$\Sigma(t)$ satisfies the following Riccati differential equation,

$$\dot{\Sigma}(t) = A(t)\Sigma(t) + \Sigma(t)A(t)^T + L(t)Q_c(t)L^T(t) - \Sigma(t)C(t)^T R_c(t)^{-1}C(t)\Sigma(t) \tag{8.19}$$

■

A block diagram of an optimal estimator or Kalmam-Bucy filter is shown in Figure 8.1. A block diagram of the covariance estimate and gain computation for the Kalman-Bucy filter is shown in Figure 8.2. There are a few interesting observations which are summarized below:

(i) As a consequence of the continuous time formulation, in the Kalman-Bucy filter, there is no prediction stage. Only one step is sufficient for the estimation of states.

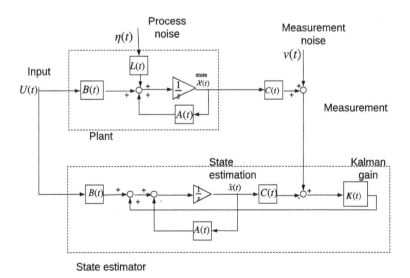

FIGURE 8.1: Block diagram of the Kalman-Bucy filter.

(ii) It may also be noted that the measurement noise covariance $R_c(t)$ has to be invertible and non zero. This means that no "perfect" measurements are allowed in the said filter. If some of the components of the measurement are accurate, then one can assume a very small noise for those components. Another approach may be to isolate those components because there is no need for estimation.

(iii) Assuming zero process noise covariance and multiplying $\Sigma(t)^{-1}$ by Eq. (8.8) from both left and right side we receive

$$\Sigma(t)^{-1}\dot{\Sigma}(t)\Sigma(t)^{-1} = \Sigma(t)^{-1}A(t) + A(t)^T\Sigma(t)^{-1} - C(t)^TR_c(t)^{-1}C(t). \tag{8.20}$$

Using the identity,

$$\frac{d}{dt}\Sigma(t)^{-1} = -\Sigma(t)^{-1}\dot{\Sigma}(t)\Sigma(t)^{-1}, \tag{8.21}$$

the above equation becomes,

$$\frac{d}{dt}\Sigma(t)^{-1} = -\Sigma(t)^{-1}A(t) - A(t)^T\Sigma(t)^{-1} + C(t)^TR_c(t)^{-1}C(t). \tag{8.22}$$

The inverse of covariance is known as the information matrix, which also follows a first order differential equation, described above. The solution

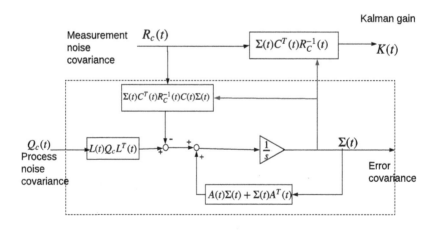

FIGURE 8.2: Block diagram of the gain and error covariance calculation of the Kalman-Bucy filter.

of the above equation can be written as

$$\Sigma(t)^{-1} = \phi(t_0, t)^T \Sigma(t_0)^{-1} \phi(t_0, t) + \int_{t_0}^t \phi(\tau, t)^T C(\tau)^T R_c(\tau)^{-1} C(\tau) \phi(\tau, t) d\tau, \tag{8.23}$$

where $\phi(\cdot)$ is the state transition matrix of $A(t)$. The last term of the above equation is called observability Gramian. From the above, it can be seen that if no prior information is available i.e., $\Sigma(t_0)^{-1}$ is singular, such information is available from the observations if the observability Gramian

$$G \triangleq \int_{t_0}^t \phi(\tau, t)^T C(\tau)^T R_c(\tau)^{-1} C(\tau) \phi(\tau, t) d\tau, \tag{8.24}$$

is positive definite. For a time invariant system, positive definiteness of the Gramian matrix is equivalent to the observability criteria. Hence, observability of a system guarantees the existence and boundedness of the error covariance matrix.

(iv) Please note that if the noises are stationary, i.e., if the noise covariances are constant, the system is linear, time is invariant, and the dynamics for $\Sigma(t)$ is stable, the observer gain converges to a constant matrix and $\Sigma(t)$ satisfies the algebraic Riccati equation,

$$A\Sigma + \Sigma A^T + LQ_cL^T - \Sigma C^T R_c^{-1} C\Sigma = 0. \tag{8.25}$$

(v) The term $(\mathcal{Y}(t) - C(t)\hat{\mathcal{X}}(t))$ is known as innovation or residue. It can be shown that the autocorrelation function of innovation is the impulse function and has a constant power spectral density. So an innovation is a white process. This property is used to validate the filter in real life applications.

Example: Consider the following system, which is an oscillator (with poles at $\pm j$)

$$\dot{\mathcal{X}}(t) = A\mathcal{X}(t) + L\eta(t) = \begin{bmatrix} 0 & 1 \\ -1 & 0 \end{bmatrix} \mathcal{X}(t) + \begin{bmatrix} 0 \\ 1 \end{bmatrix} \eta(t), \tag{8.26}$$

with measurement

$$\mathcal{Y}(t) = C\mathcal{X}(t) + v(t) = \begin{bmatrix} 1 & 0 \end{bmatrix} \mathcal{X}(t) + v(t). \tag{8.27}$$

Noises are Gaussian $\eta(t) \sim \mathcal{N}(0, Q_c)$, and $v(t) \sim \mathcal{N}(0, R_c)$, where $Q_c = 1$ and $R_c = 3$, respectively. Using Eq.(8.25), the Riccati equation of error covariance at steady state becomes

$$\begin{bmatrix} 0 & 1 \\ -1 & 0 \end{bmatrix} \Sigma + \Sigma \begin{bmatrix} 0 & -1 \\ 1 & 0 \end{bmatrix} - \frac{1}{3} \Sigma \begin{bmatrix} 1 & 0 \\ 0 & 0 \end{bmatrix} + \Sigma \begin{bmatrix} 0 & 0 \\ 0 & 1 \end{bmatrix} = 0 \tag{8.28}$$

From the above equation, we could write

$$2\Sigma_{12} - \frac{1}{3}\Sigma_{11}^2 = 0,$$

$$\Sigma_{22} - \Sigma_{11} - \frac{1}{3}\Sigma_{12}\Sigma_{11} = 0, \qquad (8.29)$$

$$-2\Sigma_{12} - \frac{1}{3}\Sigma_{12}^2 = -1.$$

The above equations are solved and we obtain $\Sigma_{12} = -3 \pm 2\sqrt{3} = 0.464$. The other two Σ becomes, $\Sigma_{11} = 1.67$ and $\Sigma_{22} = 1.93$ (we exclude the imaginary solution). Please note that the error covariance matrix is positive definite. The Kalman gain becomes

$$K = \Sigma C^T R_c^{-1} = \begin{bmatrix} 0.557 \\ 0.155 \end{bmatrix}. \qquad (8.30)$$

Thus the steady state filter equation is

$$\dot{\hat{\mathcal{X}}}(t) = \begin{bmatrix} 0 & 1 \\ -1 & 0 \end{bmatrix} \hat{\mathcal{X}}(t) + \begin{bmatrix} 0.557 \\ 0.155 \end{bmatrix} [\mathcal{Y}(t) - C\hat{\mathcal{X}}(t)]. \qquad (8.31)$$

As we know the measurement $\mathcal{Y}(t)$, the above equation can be solved with an assumed initial state. The Kalman-Bucy filter loop and flow diagram is drawn in Figure 8.3.

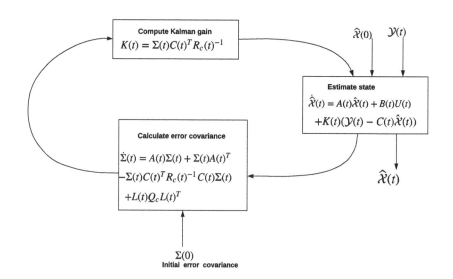

FIGURE 8.3: The Kalman-Bucy filter loop.

8.2.2 Nonlinear continuous time system

We have seen that a closed form solution for a continuous time linear Gaussian system is available. However, now we shall see how we can estimate states of a continuous time nonlinear system. A nonlinear system in general is represented by Eq. (8.3). The easiest approach is to fit the linear Kalman-Bucy filtering equations after linearizing the nonlinear equations. This is known as the extended Kalman-Bucy filter .

8.2.2.1 The extended Kalman-Bucy filter

The structure of the state estimator will be in the form of

$$\dot{\hat{\mathcal{X}}}(t) = f(\hat{\mathcal{X}}(t), \mathcal{U}(t)) + K(t)(\mathcal{Y}(t) - g(\hat{\mathcal{X}}, t)). \tag{8.32}$$

Subtracting the above equation from Eq. (8.3), we obtain,

$$\dot{\mathcal{X}}(t) - \dot{\hat{\mathcal{X}}}(t) = f(\mathcal{X}(t), \mathcal{U}(t)) - f(\hat{\mathcal{X}}(t), \mathcal{U}(t)) + L(t)\eta(t) - K(t)(\mathcal{Y}(t) - g(\hat{\mathcal{X}}, t)), \tag{8.33}$$

or,

$$\dot{e}(t) = F(\mathcal{X}(t), \hat{\mathcal{X}}(t), \mathcal{U}(t)) + L(t)\eta(t) - K(t)(g(\mathcal{X}, t) + v(t) - g(\hat{\mathcal{X}}, t)), \tag{8.34}$$

or,

$$\dot{e}(t) = F(\mathcal{X}(t), \hat{\mathcal{X}}(t), \mathcal{U}(t)) + L(t)\eta(t) - K(t)v(t) - K(t)G(\mathcal{X}(t), \hat{\mathcal{X}}(t)), \tag{8.35}$$

where $F(\mathcal{X}(t), \hat{\mathcal{X}}(t), \mathcal{U}(t)) = f(\mathcal{X}(t), \mathcal{U}(t)) - f(\hat{\mathcal{X}}(t), \mathcal{U}(t))$, and $G(\mathcal{X}(t), \hat{\mathcal{X}}(t)) = g(\mathcal{X}, t) - g(\hat{\mathcal{X}}, t)$. Note that the error dynamics is a nonlinear differential equation. Further, after linearization the error dynamics can be approximately written as

$$\dot{e}(t) \approx \frac{\partial F}{\partial e} |_{\hat{\mathcal{X}}(t)} e(t) + L(t)\eta(t) - K(t)v(t) - K(t)\frac{\partial G}{\partial e} |_{\hat{\mathcal{X}}(t)} e(t),$$
$$\approx (\tilde{A} - K(t)\tilde{C})e(t) + L(t)\eta(t) - K(t)v(t), \tag{8.36}$$

where $\frac{\partial F}{\partial e} |_{\hat{\mathcal{X}}(t)} = \tilde{A}$, and $\frac{\partial G}{\partial e} |_{\hat{\mathcal{X}}(t)} = \tilde{C}$. Following Theorem 4, the expression of error covariance equation can be derived as

$$\dot{\Sigma}(t) = \tilde{A}(t)\Sigma(t) + \Sigma(t)\tilde{A}(t)^T + L(t)Q_c(t)L(t)^T - \Sigma(t)\tilde{C}(t)^T R_c(t)^{-1}\tilde{C}(t)\Sigma(t) \tag{8.37}$$

The approach taken in the extended Kalman-Bucy filter is not optimal. Many times it leads to track loss or unacceptable estimation error, particularly for a strongly nonlinear system. Note that there is no optimal estimator available (in general) for nonlinear systems.

8.3 Continuous-discrete filtering

In the above section, we discussed continuous time filtering where the process and measurement equations are both in a continuous time domain. This section is dedicated to the continuous-discrete system where the process model is described in continuous time but the measurements remain in a discrete time domain. Such a description of a system naturally arises in practice as the processes are modeled from the physical laws and a sensor captures the measurement data at a fixed interval of time. The filtering problems of such systems are common in many real-life applications, such as target tracking [138, 122], navigation [13], control systems [154] etc.

A linear continuous-discrete system can be expressed as

$$\dot{\mathcal{X}}(t) = A(t)\mathcal{X}(t) + B(t)\mathcal{U}(t) + L(t)\eta(t), \tag{8.38}$$

and

$$\mathcal{Y}_k = C_k \mathcal{X}_k + v_k. \tag{8.39}$$

Further, we assume that the white noise $\eta(t)$ is Gaussian with zero mean and $Q_c(t)$ covariance, i.e., $\mathbb{E}[\eta(t)] = 0$, $\mathbb{E}[\eta(t)\eta(t)^T] = Q_c(t)$. The measurement noise v_k is white Gaussian with zero mean and R_k covariance, i.e., $\mathbb{E}[v_k] = 0$, $\mathbb{E}[v_k v_k^T] = R_k$. Further, the process and measurement noises are assumed as uncorrelated.

The derivation of the continuous-discrete Kalman filter is a little different from the continuous time Kalman filter because measurement is available only at discrete time-steps. Hence, the continuum time propagation model should not involve any measurement information. The estimated state equation becomes

$$\dot{\hat{\mathcal{X}}}(t) = A(t)\hat{\mathcal{X}}(t) + B(t)\mathcal{U}(t). \tag{8.40}$$

Subtracting the above equation from (8.38),

$$\dot{e}(t) = \dot{\mathcal{X}}(t) - \dot{\hat{\mathcal{X}}}(t) = A(t)e(t) + L(t)\eta(t). \tag{8.41}$$

The solution of the above equation can be written as

$$e(t) = \phi(t, t_0)e(t_0) + \int_{t_0}^{t} \phi(t, \tau)L(\tau)\eta(\tau)d\tau, \tag{8.42}$$

where $\phi(t, t_0)$ is the state transition matrix of $A(t)$. The dynamics of error covariance can be derived as follows:

$$\begin{aligned} \dot{\Sigma}(t) &= \mathbb{E}[\dot{e}(t)e(t)^T + e(t)\dot{e}(t)^T] \\ &= \mathbb{E}[\dot{e}(t)e(t)^T] + (\mathbb{E}[\dot{e}(t)e(t)^T])^T \end{aligned} \tag{8.43}$$

Now,

$$
\begin{aligned}
\mathbb{E}[\dot{e}(t)e(t)^T] &= \mathbb{E}[(A(t)e(t) + L(t)\eta(t))e(t)^T], \\
&= A(t)\mathbb{E}[e(t)e(t)^T] + L(t)\mathbb{E}[\eta(t)e(t)^T] \\
&= A(t)\Sigma(t) + L(t)\mathbb{E}[\eta(t)e(t)^T].
\end{aligned} \tag{8.44}
$$

Using (8.42), the second term of the above equation becomes,

$$
\begin{aligned}
L(t)\mathbb{E}[\eta(t)e(t)^T] &= L(t)\mathbb{E}[\eta(t)e(t_0)\phi^T(t,t_0)] \\
&\quad + L(t)\mathbb{E}[\int_{t_0}^{t} \eta(t)\eta(\tau)^T L(\tau)^T \phi(t,\tau)^T d\tau], \\
&= L(t)\int_{t_0}^{t} Q_c(t)\delta(t-\tau)L(\tau)^T \phi(t,\tau)^T d\tau \\
&= 1/2 L(t)Q_c(t)L(t)^T.
\end{aligned} \tag{8.45}
$$

Finally Eq. (8.43) becomes

$$
\dot{\Sigma}(t) = A(t)\Sigma(t) + \Sigma(t)A(t)^T + L(t)Q_c(t)L(t)^T \tag{8.46}
$$

In continuous-discrete Kalman filtering, the calculation of the Kalman gain and the update stage are carried out with the discrete Kalman filter equations whereas the measurement update stage is carried out with the solution of Eqs.(8.40) and (8.46). The algorithm for the continuous-discrete Kalman-Bucy filter is summarized in Algorithm 10.

If the continuous-discrete system is nonlinear in nature, the extended Kalman-Bucy filter algorithm can be adopted. A nonlinear continuous-discrete system can be expressed in states space as follows:

$$
\dot{\mathcal{X}}(t) = f(\mathcal{X}(t)) + B(t)\mathcal{U}(t) + L(t)\eta(t), \tag{8.47}
$$

and

$$
\mathcal{Y}_k = \gamma_k(\mathcal{X}_k) + v_k, \tag{8.48}
$$

where $f(\cdot)$, and $\gamma_k(\cdot)$ are nonlinear functions. The sampling period is $T = t_k - t_{k-1}$ and considered as constant. Following Eq. (8.40), the state prediction equation of continuous-discrete EKF (CD-EKF) can be written as,

$$
\dot{\hat{\mathcal{X}}}(t) = F_k\hat{\mathcal{X}}(t) + B(t)\mathcal{U}(t), \tag{8.49}
$$

where the initial condition is $\hat{\mathcal{X}}(t_k) = \hat{\mathcal{X}}_{k|k}$, and $F_k = \frac{\partial f(\cdot)}{\partial \mathcal{X}}|_{\hat{\mathcal{X}}_{k|k}}$, and the integral has to be performed on time duration of T, which is up to time t_{k+1}. After solving the above differential equation with the initial condition, we receive $\hat{\mathcal{X}}_{k+1|k}$. Similar to (8.46) the error covariance update equation can be written as

$$
\dot{\Sigma}(t) = F_k\Sigma(t) + \Sigma(t)F_k^T + L(t)Q_c(t)L(t)^T. \tag{8.50}
$$

Algorithm 10 Continuous-discrete Kalman-Bucy filter

$$[\hat{\mathcal{X}}_{k|k}, \Sigma_{k|k}] = \text{CDKB}[\mathcal{Y}_k]$$

- Initialize with $\hat{\mathcal{X}}_{0|0}$ and $\Sigma_{0|0}$

- For $k = 1 : N$

  Time update

  - $\hat{\mathcal{X}}_{k+1|k}$ can be obtained by solving Eq. (8.40)

  $$\hat{\mathcal{X}}_{k+1|k} = \phi(t_{k+1}, t_k)\hat{\mathcal{X}}_{k|k} + \int_{t_k}^{t_{k+1}} \phi(t_{k+1} - \tau)B(\tau)\mathcal{U}(\tau)d\tau$$

  - $\Sigma_{k+1|k}$ can be obtained by solving Eq. (8.46)
  $$\Sigma_{k+1|k} = \phi(t_{k+1}, t_k)^T \Sigma_{k|k}\phi(t_{k+1}, t_k) +$$
  $$\int_{t_k}^{t_{k+1}} \phi(t_{k+1} - \tau)^T L(\tau)Q_c(\tau)L(\tau)^T \phi(t_{k+1} - \tau)d\tau .$$

  Measurement update

  - Calculate the Kalman gain, $K_k = \Sigma_{k|k-1}C_k^T[C_k\Sigma_{k|k-1}C_k^T + R_k]^{-1}$.

  - Evaluate posterior estimate, $\hat{\mathcal{X}}_{k|k} = \hat{\mathcal{X}}_{k|k-1} + K_k[\mathcal{Y}_k - C_k\hat{\mathcal{X}}_{k|k-1}]$.

  - Evaluate posterior error covariance $\Sigma_{k|k} = (I - K_kC_k)\Sigma_{k|k-1}$.

- End for

The above equation can also be solved with the initial condition, $\Sigma(t_k) = \Sigma_{k|k}$, from t_k to t_{k+1}. Once we solve the above two equations, we obtain $\hat{\mathcal{X}}_{k+1|k}$, and $\Sigma_{k+1|k}$ which will eventually go to the measurement update step which is similar to the CD Kalman filter except the fact that the measurement matrix is replaced by its Jacobian, i.e., $C_k = \frac{\partial \gamma_k(\cdot)}{\partial \mathcal{X}}|_{\hat{\mathcal{X}}_{k+1|k}}$. The estimation process described above is summarized in Algorithm 11.

Algorithm 11 Continuous-discrete extended Kalman filter

$$[\hat{\mathcal{X}}_{k|k}, \Sigma_{k|k}] = \texttt{CDEKF}[\mathcal{Y}_k]$$

- Initialize with $\hat{\mathcal{X}}_{0|0}$ and $\Sigma_{0|0}$

- For $k = 1 : N$

  Time update

  - The Jacobian of process function $F_k = \frac{\partial f(\cdot)}{\partial \mathcal{X}}|_{\hat{\mathcal{X}}_{k|k}}$

  - Solve Eq. (8.49), with the initial condition $\hat{\mathcal{X}}(t_k) = \hat{\mathcal{X}}_{k|k}$ to obtain $\hat{\mathcal{X}}_{k+1|k}$.

  - Solve Eq. (8.50) with the initial condition $\Sigma(t_k) = \Sigma_{k|k}$, to obtain $\Sigma_{k+1|k}$.

  Measurement update

  - The Jacobian of measurement function $C_k = \frac{\partial \gamma_k(\cdot)}{\partial \mathcal{X}}|_{\hat{\mathcal{X}}_{k|k-1}}$

  - Calculate the Kalman gain, $K_k = \Sigma_{k|k-1}C_k^T[C_k\Sigma_{k|k-1}C_k^T + R_k]^{-1}$.

  - Evaluate posterior estimate, $\hat{\mathcal{X}}_{k|k} = \hat{\mathcal{X}}_{k|k-1} + K_k[\mathcal{Y}_k - C_k\hat{\mathcal{X}}_{k|k-1}]$.

  - Evaluate posterior error covariance $\Sigma_{k|k} = (I - K_kC_k)\Sigma_{k|k-1}$.

8.3.1 Nonlinear continuous time process model

Let us consider a nonlinear stochastic dynamic system represented by the continuous process model as

$$d\mathcal{X}(t) = f(\mathcal{X}(t), t)dt + \sqrt{Q}d\beta(t). \tag{8.51}$$

For continuous systems, the prior density is obtained from the posterior using the Fokker-Planck equation, also known as Kolmogorov's forward equation [47, 7]. This can be defined as

$$\frac{\partial p(\mathcal{X}_t|\mathcal{Y}_{1:k})}{\partial t} = -f^T\frac{\partial p(\mathcal{X}_t|\mathcal{Y}_{1:k})}{\partial \mathcal{X}_t} - tr\left(\frac{\partial f}{\partial t}\right) + \frac{1}{2}tr(Q)\frac{\partial^2 p(\mathcal{X}_t|\mathcal{Y}_{1:k})}{\partial \mathcal{X}_t^2}, \tag{8.52}$$

where $p(\mathcal{X}_t|\mathcal{Y}_{1:k})$ represents the posterior density, and $tr(\cdot)$ denotes the trace of a matrix. The prior probability density can be obtained by solving the above equation. The existence of exact solutions for the Fokker-Planck equation, given in Eq. (8.52), is restricted to a very few cases such as for the formulation of the Kalman-Bucy filter [90], Beneš filter [14] and in the work proposed by Daum [38] etc. Apart from these cases, Eq. (8.52) has to be solved using approximate measures.

8.3.2   Discretization of process model using Runge-Kutta method

The continuous process Eq. (8.51) can be solved by discretizing it on much smaller time step compared to the sampling time of a measurement sensor. Consider the notation $\mathcal{X}_k^j$ for $\mathcal{X}(t)$ at time $t = t_k + j\delta$, with $1 \leq j \leq m_{step}$ and $\delta = T/m_{step}$, where T is the sampling time of the measurement device.

The mean $(\hat{\mathcal{X}}(t_k + n\delta))$ of the state at an intermediate time step, $t_k + n\delta$ is recursively approximated by an ith order explicit Runge-Kutta method given by

$$\hat{\mathcal{X}}_k^j = \hat{\mathcal{X}}_k^{j-1} + (\mathfrak{R}_i^\delta f)(\hat{\mathcal{X}}_k^{j-1}, t_k + j\delta), \tag{8.53}$$

where $\mathfrak{R}_i^\delta$ is an ith order Runge-Kutta operator. In the sense of EKF, the error covariance $(\Sigma(t_k + n\delta))$ of intermediate time step, $t_k + n\delta$ could be written as

$$\Sigma_k^j = \Sigma_k^{j-1} + (\mathfrak{R}_i^\delta r)(\Sigma_k^{j-1}, t_k + j\delta), \tag{8.54}$$

where $r(\cdot)$ is a function defined by $F_k\Sigma_k + \Sigma_k F_k^T + L_k Q_c L_k^T$, and F_k is the Jacobian matrix. The Runge-Kutta operator $\mathfrak{R}_i^\delta$ is defined in Table 8.1 [46]. The prior mean and covariance at the $(k + 1)^{th}$ step are $\hat{\mathcal{X}}_{k+1|k} = \hat{\mathcal{X}}_k^{m_{step}}$ and $\Sigma_{k+1|k} = \Sigma_k^{m_{step}}$ respectively. After calculating the prior density, the posterior density is obtained using the Bayes' rule, which is the same as that of a conventional discrete-discrete filter.

The Runge-Kutta method is very popular among practitioners. Step size δ is recommended to be as small as possible. For large δ, stability of the scheme is not guaranteed. Further, in such a case the approximate solution of the covariance matrix can lead to a negative semi definite matrix. The above discretization method is described in the sense of CD-EKF. The Runge-Kutta method described above can also be utilized to implement the continuous-discrete UKF (CD-UKF) and other CD deterministic sample point filters. We shall discuss this in the next section.

8.3.3   Discretization using Ito-Taylor expansion of order 1.5

We have seen that the process update can be done with the help of a Runge-Kutta method. Another method which is very popular in the filtering community is Ito-Taylor expansion of order 1.5 [136, 97]. In this subsection, we shall discuss it. Discretization using the Ito-Taylor expansion of order 1.5

TABLE 8.1: Runge-Kutta operator of different orders.

Euler $(i = 1)$	$\mathfrak{R}_1^\delta f(\mathcal{X}, t) = \delta f(\mathcal{X}, t)$
Heun $(i = 2)$	$\mathfrak{R}_2^\delta f(\mathcal{X}, t) = \frac{\delta}{2}(k1 + k2)$
	$k1 = f(\mathcal{X}, t)$
	$k2 = f(\mathcal{X} + \delta k1, t + \delta)$
RK4 $(i = 4)$	$\mathfrak{R}_4^\delta f(\mathcal{X}, t) = \frac{\delta}{6}(k1 + 2k2 + 2k3 + k4)$
	$k1 = f(\mathcal{X}, t)$
	$k2 = f(\mathcal{X} + \frac{\delta}{2}k1, t + \frac{\delta}{2})$
	$k3 = f(\mathcal{X} + \frac{\delta}{2}k2, t + \frac{\delta}{2})$
	$k4 = f(\mathcal{X} + \delta k3, t)$

is more accurate than the Euler Maruyama method (often called simple Euler method), mentioned in Table 8.1, since it represents the process by considering higher order terms.

Applying the Ito-Taylor expansion of order 1.5 to Eq. (8.51) for the time interval $(t, t + \delta)$ [164], we get

$$\mathcal{X}(t + \delta) = \mathcal{X}(t) + \delta f(\mathcal{X}(t), t) + \frac{1}{2}\delta^2(\mathbb{L}_0 f(\mathcal{X}(t), t)) + \sqrt{Q}\beta + (\mathbb{L}f(\mathcal{X}(t), t))\beta',$$
$$(8.55)$$

where

$$\mathbb{L}_0 = \frac{\partial}{\partial t} + \sum_{i=1}^n f_i \frac{\partial}{\partial \mathcal{X}_i} + \frac{1}{2} \sum_{j,p,q=1}^n \sqrt{Q}_{p,j}\sqrt{Q}_{q,j} \frac{\partial^2}{\partial \mathcal{X}_p \partial \mathcal{X}_q},$$

and

$$\mathbb{L} = \sum_{j,i=1}^n \sqrt{Q}_{i,j} \frac{\partial}{\partial \mathcal{X}_i}.$$

Here, $\sqrt{Q}_{i,j}$ represents the square-root of the $(i, j)^{th}$ element of Q. The deterministic part of Eq. (8.51) can be represented as

$$f_d(\mathcal{X}(t), t) = \mathcal{X}(t) + \delta f(\mathcal{X}(t), t) + \frac{1}{2}\delta^2(\mathbb{L}_0 f(\mathcal{X}(t), t)).$$

β, β' are n dimensional Gaussian random variables which are correlated and independent of $\mathcal{X}(t)$. They are generated from a pair of independent n dimensional standard normal random variables (β_1, β_2) as $\beta = \sqrt{\delta}\beta_1$ and $\beta' = \frac{1}{2}\delta^{3/2}(\beta_1 + \frac{\beta_2}{\sqrt{3}})$. Their associated covariances can be written as $\mathbb{E}[\beta\beta^T] = \delta I_n$, $\mathbb{E}[\beta\beta'^T] = \frac{1}{2}\delta^2 I_n$ and $\mathbb{E}[\beta'\beta'^T] = \frac{1}{3}\delta^3 I_n$.

8.3.4 Continuous-discrete filter with deterministic sample points

As has been discussed in the previous chapters, the deterministic sample point filters or Gaussian filters approximate the posterior and prior pdfs as Gaussian and are characterized by propagation of the mean and the covariance.

The deterministic sample point filters or Gaussian filters approximate the posterior and prior pdfs as Gaussian and are characterized with mean and covariance. The prior mean and the prior covariance at each jth step of discretization can be defined as

$$\hat{\mathcal{X}}_k^{j+1} = \mathbb{E}[\mathcal{X}_k^{j+1}] = \mathbb{E}[f_d(\mathcal{X}_k^j)] \qquad (8.56)$$

and

$$
\begin{aligned}
\Sigma_k^{j+1} &= \mathbb{E}[(\mathcal{X}_k^{j+1} - \hat{\mathcal{X}}_k^{j+1})(\mathcal{X}_k^{j+1} - \hat{\mathcal{X}}_k^{j+1})^T] \\
&= \mathbb{E}[(f_d(\mathcal{X}_k^j) + \sqrt{Q}\beta + \mathbb{L}f(\mathcal{X}_k^j)\beta' - \hat{\mathcal{X}}_k^{j+1})(f_d(\mathcal{X}_k^j) + \sqrt{Q}\beta \\
&\quad + \mathbb{L}f(\mathcal{X}_k^j)\beta' - \hat{\mathcal{X}}_k^{j+1})^T] \\
&= \mathbb{E}[f_d(\mathcal{X}_k^j)f_d(\mathcal{X}_k^j)^T] - \hat{\mathcal{X}}_k^{j+1}\hat{\mathcal{X}}_k^{j+1\,T} + \delta Q \\
&\quad + \frac{\delta^2}{2}[\sqrt{Q}\mathbb{L}f(\hat{\mathcal{X}}_k^j)^T + \mathbb{L}f(\hat{\mathcal{X}}_k^j)\sqrt{Q}^T] + \frac{\delta^3}{3}\mathbb{L}f(\hat{\mathcal{X}}_k^j)\mathbb{L}f(\hat{\mathcal{X}}_k^j)^T.
\end{aligned}
\qquad (8.57)
$$

Here, we assume $\mathbb{L}f(\mathcal{X}_k^j) \approx \mathbb{L}f(\hat{\mathcal{X}}_k^j)$ to simplify the covariance calculation [7]. After the time update, the prior mean and covariance at the $(k+1)^{th}$ step are $\hat{\mathcal{X}}_{k+1|k} = \hat{\mathcal{X}}_k^{m_{step}}$ and $\Sigma_{k+1|k} = s_k^{m_{step}}$ respectively. After calculating the prior density, the posterior density is obtained using the Bayes' rule, which is the same as that of a conventional discrete-discrete filter.

It should be noted that for more accurate prediction using the process model a higher m should be selected. For $m = 1$, the continuous-discrete filtering algorithm reduces to ordinary discrete time filtering.

Now, with the discretization procedure described above, it will be easy to formulate continuous-discrete Gaussian filters such as CD-UKF [97], CD-CKF [7], CD-GHF [6] etc. We know that, in a Gaussian filter or a deterministic sample point filter, the prior and the posterior pdfs are approximated with the Gaussian pdf as represented by support points and weights. To implement CD Gaussian filters, initially the points and weights are generated for a zero mean unity variance process. During the time update, they are spread with the help of posterior mean and covariance. Next the support points are propagated with the process equation. The mean is calculated as the weighted sum of the propagated support points and covariance is calculated around the mean using the formula described above. The algorithm of the continuous-discrete Gaussian filter is described in Algorithm 12.

Algorithm 12 CD deterministic sample point filter

$$[\hat{\mathcal{X}}_{k+1|k+1}, \Sigma_{k+1|k+1}] = \text{CDCQKF}[\mathcal{Y}_{k+1}]$$

- Initialize with $\hat{\mathcal{X}}_{0|0}$ and $\Sigma_{0|0}$

- Calculate support points ξ_i and w_i for $i = 1, \cdots, n_p$

- For $k = 1 : N$

  Time update

  - $\hat{\mathcal{X}}^1_{k|k} = \hat{\mathcal{X}}_{k|k}$ and $\Sigma^1_{k|k} = \Sigma_{k|k}$

- For $j = 1 : m_{step}$
  - Cholesky decomposition of posterior error covariance, $\Sigma^j_{k|k} = S^j_{k|k} S^{j\ T}_{k|k}$.
  - Spread the CQ points, $\chi^j_{i,k} = \hat{\mathcal{X}}^j_{k|k} + S^j_{k|k}\xi_i$,

  - $\chi^{j+1}_{i,k|k} = f_d(\chi^j_{i,k})$

  - $\hat{\mathcal{X}}^{j+1}_{k|k} = \sum_{i=1}^{n_p} w_i \chi^{j+1}_{i,k|k}$

  - $\Sigma^{j+1}_{k|k} = \sum_{i=1}^{n_p} w_i [\chi^{j+1}_{i,k|k} - \hat{\mathcal{X}}^{j+1}_{k|k}][\chi^{j+1}_{i,k|k} - \hat{\mathcal{X}}^{j+1}_{k|k}]^T - \hat{\mathcal{X}}^{j+1}_{k|k}\hat{\mathcal{X}}^{j+1\ T}_{k|k} + \delta Q$
    $+ \frac{\delta^2}{2}[\sqrt{Q}\mathbb{L}f(\hat{\mathcal{X}}^j_{k|k})^T + \mathbb{L}f(\hat{\mathcal{X}}^j_{k|k})\sqrt{Q}^T] + \frac{\delta^3}{3}\mathbb{L}f(\hat{\mathcal{X}}^j_{k|k})\mathbb{L}f(\hat{\mathcal{X}}^j_{k|k})^T$

- End for
  - $\hat{\mathcal{X}}_{k+1|k} = \hat{\mathcal{X}}^{j+1}_{k|k}$

  - $\Sigma_{k+1|k} = \Sigma^{j+1}_{k|k}$

  Measurement update

  - Perform Cholesky decomposition $\Sigma_{k+1|k} = S_{k+1|k}S^T_{k+1|k}$.

  - $\chi_{i,k+1} = \hat{\mathcal{X}}_{k+1|k} + S_{k+1|k}\xi_i$,

  - $Y^*_{i,k+1} = \gamma(\chi_{i,k+1})$

  - $\hat{\mathcal{Y}}_{k+1} = \sum_{i=1}^{n_p} w_i Y^*_{i,k+1}$

  - $\Sigma_{\mathcal{Y},k+1} = \sum_{i=1}^{n_p} w_i [Y^*_{i,k+1} - \hat{\mathcal{Y}}_{k+1}][Y^*_{i,k+1} - \hat{\mathcal{Y}}_{k+1}]^T + R_k$

  - $\Sigma_{\mathcal{XY},k+1} = \sum_{i=1}^{n_p} w_i [\chi_{i,k+1} - \hat{\mathcal{X}}_{k+1|k}][Y^*_{i,k+1} - \hat{\mathcal{Y}}_{k+1}]^T$

  - Kalman gain, $K = \Sigma_{\mathcal{XY},k+1}\Sigma^{-1}_{\mathcal{Y},k+1}$

  - Posterior estimate $\hat{\mathcal{X}}_{k+1|k+1} = \hat{\mathcal{X}}_{k+1|k} + K[\mathcal{Y}_{k+1} - \hat{\mathcal{Y}}_{k+1}]$

  - Posterior error covariance $\Sigma_{k+1|k+1} = \Sigma_{k+1|k} - K\Sigma_{\mathcal{XY},k+1}K^T$

- End for

8.4   Simulation examples

In this section, continuous-discrete $4n + 1$ sigma point filter (as discussed in Section 3.4.2) which we call by the name continuous-discrete new unscented Kalman filter (CD-NUKF), and continuous-discrete sparse Gauss-Hermite filter (CD-SGHF) have been implemented on a few test problems. The performance of the filters is studied and compared with that of CD-UKF and CD-CQKF. For discrete-discrete UKF, CQKF and SGHF, please see earlier chapters.

8.4.1   Single dimensional filtering problem

Here we consider a single dimensional system with the process model described by (8.51) with $f(\mathcal{X}(t)) = 5\mathcal{X}(1 - \mathcal{X}^2)$ and $Q = b^2 T$. The measurement model is described as $\mathcal{Y}_k = \gamma(\mathcal{X}_k) + v_k$, where $\gamma(\mathcal{X}_k) = T\mathcal{X}_k(1 - 0.5\mathcal{X}_k)$ and $v_k \sim \mathcal{N}(0, d^2 T)$. The sampling time $T = 0.01s$ and the simulations are done for a total of 4 seconds. The value of simulation parameters b and d are taken as 0.5 and 0.05, respectively.

Now the corresponding stochastic difference equation can be derived as

$$\mathcal{X}_k^{j+1} = f_d(\mathcal{X}_k^j) + \sqrt{Q}\eta + (\mathbb{L}f(\mathcal{X}_k^j))\eta', \qquad (8.58)$$

where $\mathcal{X}_k^j$ denotes $\mathcal{X}(t)$ at time $t = t_k + j\delta$, with $1 \le j \le m_{step}$ and $\delta = T/m_{step}$. The noise free process function can be given as

$$f_d(\mathcal{X}_k^j) = \mathcal{X}_k^j + \delta f(\mathcal{X}_k^j) + \frac{\delta^2}{2}\left(f(\mathcal{X}_k^j)(5 - 15\mathcal{X}_k^{j^2}) - \frac{Q}{2}30\mathcal{X}_k^j\right),$$

and $\mathbb{L}f(\mathcal{X}_k^j) = \sqrt{Q}(5 - 15\mathcal{X}_k^{j^2})$. All the filters were initialized with an estimate $\hat{\mathcal{X}}_{0|0} = -0.8$ and initial error covariance $\Sigma_{0|0} = 2$, with the truth being initialized as $\mathcal{X}_{0|0} = -0.2$. The accuracy level of CD-CQKF and CD-SGHF is taken as $n' = L = 2$ such that a fair comparison with CD-NUKF is achieved. Filtering performance has been compared in terms of RMSE and the number of track-loss.

Figure 8.4 shows the RMSE out of 500 Monte Carlo runs for CD-UKF, CD-CQKF, CD-NUKF and CD-SGHF, with $m_{step} = 10$. From this figure, it can be observed that the proposed filter CD-NUKF performs with better accuracy. Filtering performance has also been studied in terms of the number of diverged tracks. Track-loss is the number of cases when the estimation error goes beyond a pre-defined value, and here we took it as $|X_{k_{max}} - \hat{X}_{k_{max}}| > 1$, where k_{max} is the final time step. The number of track-loss in percentage for different filters is summarized in Table 8.2 and we can observe the superior filtering accuracy of CD-NUKF, with zero number of diverged tracks.

TABLE 8.2: Problem 1: % of track-loss and number of points.

Filter	Track-loss (%)	No. of points
CD-UKF	0.2	3
CD-CQKF	0.2	4
CD-SGHF	0.2	5
CD-NUKF	0	5

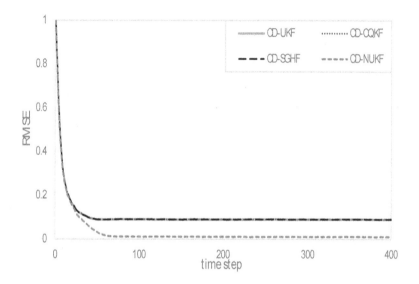

FIGURE 8.4: Problem 1: RMSE of the single dimensional problem.

8.4.2 Estimation of harmonics

In this section, CD filters are implemented for estimating the frequency and amplitude of three superimposed harmonic signals. The process and measurement models are considered the same as that given in Section 5.6.4. The error covariance for state variables f_i and a_i, are taken as $\sigma_{f_i}^2 = 0.1 \times 10^{-6}$ and $\sigma_{a_i}^2 = 0.5 \times 10^{-6}$, respectively for $i = 1, 2, 3$. To define the measurement noise covariance, R, covariance of individual measured signals are assumed as $\sigma_n^2 = 0.09$. Sampling frequency of the signal is set to be $5000Hz$. Since the process model follows a random walk, the drift function $F(\mathcal{X}(t)) = 0$. Hence the time update can be defined as

$$\hat{\mathcal{X}}_{k|k}^{j+1} = F\hat{\mathcal{X}}_k^j,$$
$$\Sigma_{k|k}^{j+1} = F\Sigma_{k|k}^j F^T + \delta Q,$$

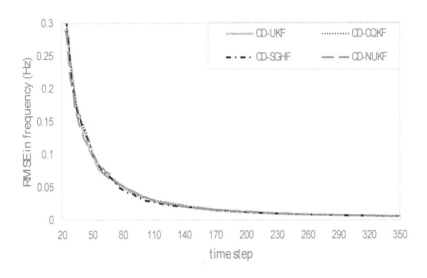

FIGURE 8.5: Problem 2: RMSE of frequency in Hz.

where $F = I_n$ and the prior mean and covariance can be identified as $\hat{\mathcal{X}}_{k+1|k} = \hat{\mathcal{X}}_{k|k}^{m_{step}}$ and $\Sigma_{k+1|k} = \Sigma_{k|k}^{m_{step}}$.

The accuracy level for CQKF and SGHF was taken as $n' = L = 2$. The filter was initialized with an initial estimate, $\hat{\mathcal{X}}_{0|0}$, which follows normal distribution with mean $\mathcal{X}_{0|0}$ and covariance $\Sigma_{0|0}$, i.e., $\hat{\mathcal{X}}_{0|0} \sim \mathcal{N}(\mathcal{X}_{0|0}, \Sigma_{0|0})$, where $\mathcal{X}_{0|0} = [100\ 1000\ 2000\ 5\ 4\ 3]^T$ and $\Sigma_{0|0} = diag([50^2\ 50^2\ 50^2\ 4\ 4\ 4])$. The states are estimated for 350 time steps and 100 Monte Carlo runs were done for ensuring the result, with $m_{step} = 10$. Validation of results was done by comparing the RMSE of frequencies and amplitudes. The relation for calculating the RMSE in frequency and amplitude is the same as that given in Section 5.6.4.

Performance comparison of the filters is also done by finding whether any track divergence had occurred or not. In the case of frequencies, a track is considered divergent when the error between truth and estimate is above $5Hz$. For amplitude, a track divergence is defined when error is above $0.5V$. Figures 8.5 and 8.6 respectively show the RMSE in frequency and amplitude for all the filters. Out of all the filters, CD-NUKF converges faster in the case of amplitude while for frequency, it gave comparable results with other CD filters. Coming to track-loss counts, it was found that none of the filters incurred even a single diverged track. Hence for this problem, CD-NUKF gave comparable or similar filtering performance with respect to CD-UKF, CD-CQKF and CD-SGHF.

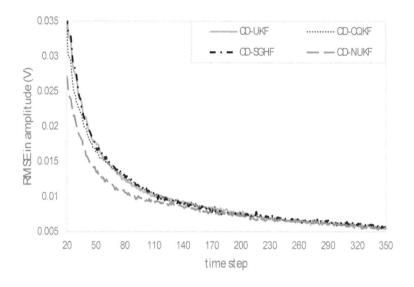

FIGURE 8.6: Problem 2: RMSE of amplitude in V.

8.4.3   RADAR target tracking problem

The 7D RADAR target tracking problem mentioned in Section 6.5.2 is considered again. Here, the state vector $\mathcal{X} = [x \ v_x \ y \ v_y \ z \ v_z \ \Omega]^T$, where (x, y, z) are the positions, (v_x, v_y, v_z) are the velocities and Ω is the the turn rate. The drift function $f(\mathcal{X}(t)) = [v_x \ -\Omega v_y \ v_y \ \Omega v_x \ v_z \ 0 \ 0]^T$ and gain matrix of the process noise is taken as $Q = diag([0, \sigma_1^2, 0, \sigma_2^2, 0, \sigma_1^2, \sigma_2^2])$. It is assumed that the measurements received are range, azimuth and elevation from a RADAR located at the origin. The noise corrupting these measurements are characterized with a covariance matrix $R = diag([\sigma_3^2, \sigma_4^2, \sigma_5^2])$. Applying the Ito-Taylor expansion of order 1.5, the resulting stochastic difference equation can be expressed as Eq. (8.58), where

$$f_d(\mathcal{X}_k^j) = \begin{bmatrix} x_k^j + \delta v_{x,k}^j - \frac{\delta^2}{2}\Omega_k^j v_{y,k}^j \\ v_{x,k}^j - \delta\Omega_k^j v_{y,k}^j - \frac{\delta^2}{2}(\Omega_k^{2\ j} v_{x,k}^j + \frac{\sigma_1\sigma_2}{2}) \\ y_k^j + \delta v_{y,k}^j + \frac{\delta^2}{2}\Omega v_{x,k}^j \\ v_{y,k}^j + \delta\Omega v_{x,k}^j - \frac{\delta^2}{2}(\Omega_k^{2\ j} v_{y,k}^j - \frac{\sigma_1\sigma_2}{2}) \\ z_k^j + \delta v_{z,k}^j \\ v_{z,k}^j \\ \Omega_k^j \end{bmatrix}$$

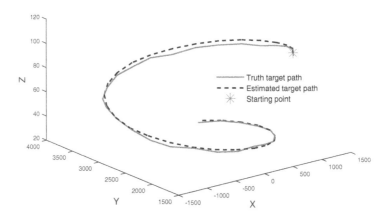

FIGURE 8.7: Problem 3: Trajectory estimated by CD-NUKF from $90s$ to $150s$.

and

$$\mathbb{L}(f(\mathcal{X}_k^j)) = \begin{bmatrix} 0 & \sigma_1 & 0 & 0 & 0 & 0 & 0 \\ 0 & 0 & 0 & -\sigma_1\Omega_k^j & 0 & 0 & -\sigma_2 v_{y,k}^j \\ 0 & 0 & 0 & \sigma_1 & 0 & 0 & 0 \\ 0 & \sigma_1\Omega_k^j & 0 & 0 & 0 & 0 & -\sigma_2 v_{x,k}^j \\ 0 & 0 & 0 & 0 & 0 & \sigma_1 & 0 \\ 0 & 0 & 0 & 0 & 0 & 0 & 0 \\ 0 & 0 & 0 & 0 & 0 & 0 & 0 \end{bmatrix},$$

where $j = 1, \cdots, m_{step}$.

For simulation purposes, the initial true state is taken as $\mathcal{X}_{0|0} = [1000m \ 0 \ 2650m \ 150m/s \ 200m/s \ 0 \ \Omega^o/s]^T$ and initial error covariance $\Sigma_{0|0} = diag([100m^2 \ 10(m/s)^2 \ 100m^2 \ 10(m/s)^2 \ 100m^2 \ 10(m/s)^2 \ (1^o/s)^2])$. The parameters $\sigma_1 = \sqrt{0.2}m/s$, $\sigma_2 = 7 \times 10^{-3} \ ^o/s$, $\sigma_3 = 50m$, $\sigma_4 = 0.1^o$ and $\sigma_5 = 0.1^o$. The sampling time is taken as $T = 2s$ and the total observation is carried out for a period of $150s$, where $j = 1, \cdots, 8$. The initial estimate $\hat{\mathcal{X}}_{0|0}$ fed to the filters are assumed to be normally distributed with mean $\mathcal{X}_{0|0}$ and covariance $\Sigma_{0|0}$. The accuracy level for CD-CQKF is $n' = 2$ and for CD-SGHF is $L = 3$. A MATLAB code for CD-NUKF is provided in Listing 8.1.

```
1 % Estimation of state for a 7D RADAR target tracking problem
2 % Implementation of NUKF
3 % Code for calculation of RMSE
4 % Code credit: Rahul Radhakrishnan, IIT Patna
5 %
6 clc;clear;randn('seed',2);
```

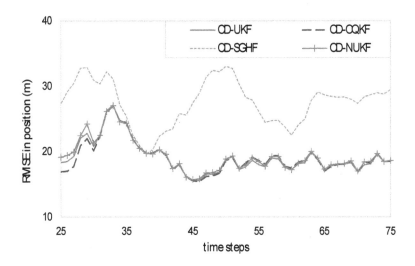

FIGURE 8.8: Problem 3: RMSE of position in m.

```
 7 n=7; % Dimension of the system
 8 sigma_1=sqrt(0.2);sigma_2=7*10^(-3);sigma_r=50;sigma_theta=0.1*
     pi/180;
 9 sigma_phi=0.1*pi/180;omega=3*pi/180;
10 Q=diag([0 sigma_1^2 0 sigma_1^2 0 sigma_1^2 sigma_2^2])/10;
11 R=diag([sigma_r^2 sigma_theta^2 sigma_phi^2]); % Meas noise cov
12 T=2;m_step=8;del=T/m_step; % del=granular step size to solve ODE
13 kmax=75;M=50; TL=0; % TL=track loss count
14 for m=1:M      % start of MC loop
15    m
16 X(:,1)=[1000;0;2650;150;200;0;omega]; % state initialization
17 y(:,1)=[sqrt(X(1,1)^2+X(3,1)^2+X(5,1)^2); atan(X(3,1)/X(1,1));
     ...
18    atan(X(5,1)/sqrt(X(1,1)^2+X(3,1)^2))]+mvnrnd(zeros(3,1),R)';
19 for k=1:kmax  % Time horizon loop
20    % -- Generation of truth state and measurements------------
21    x(:,1)=X(:,k);
22    for j=1:m_step  % ODE time step
23       fd1=x(1,j)+del*x(2,j)-del^2/2*x(7,j)*x(4,j);
24       fd2=x(2,j)-del*x(7,j)*x(4,j)-del^2/2*(x(7,j)^2*x(2,j)+...
25          (sigma_1*sigma_2/2));
26       fd3=x(3,j)+del*x(4,j)+del^2/2*x(7,j)*x(2,j);
27       fd4=x(4,j)+del*x(7,j)*x(2,j)-del^2/2*(x(7,j)^2*x(4,j)-...
28          (sigma_1*sigma_2/2));
29       fd5=x(5,j)+del*x(6,j); fd6=x(6,j); fd7=x(7,j);
30       fd=[fd1;fd2;fd3;fd4;fd5;fd6;fd7];  % Fd matrix
31       L_f(1,2)=sigma_1;L_f(2,4)=-sigma_1*x(7,j);L_f(2,7)=-
          sigma_2*x(4,j);
32       L_f(3,4)=sigma_1;L_f(4,2)=sigma_1*x(7,j);L_f(4,7)=-
          sigma_2*x(2,j);
33       L_f(5,6)=sigma_1; L_f(7,7)=0;  % L matrix
34       u1=mvnrnd(zeros(7,1),eye(7))';
```

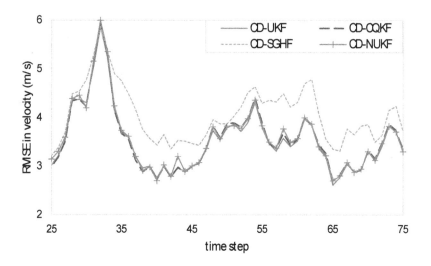

FIGURE 8.9: Problem 3: RMSE of velocity in m/s.

```
35          u2=mvnrnd(zeros(7,1),eye(7))';
36        x(:,j+1)=fd+sqrt(Q)*sqrt(del)*u1+L_f*del^(1.5)/2*(u1+u2/
              sqrt(3));
37     end
38     X(:,k+1)=x(:,j+1);  % Truth state at each step k
39     y(:,k+1)=[sqrt(X(1,k+1)^2+X(3,k+1)^2+X(5,k+1)^2); ...
40        atan(X(3,k+1)/X(1,k+1)); ...
41        atan(X(5,k+1)/sqrt(X(1,k+1)^2+X(3,k+1)^2))]+mvnrnd(zeros
              (3,1),R)';
42 end
43 P=diag([10 1 10 1 10 1 0.1*pi/180])*10; % initial error cov
44 xes(:,1)=mvnrnd(X(:,1),P)';  % Filter state initialization
45 %——————————Code for CD NUKF starts here——————————
46 for k=1:kmax    % Start of filter horizontal loop
47     xes1(:,1)=xes(:,k);
48     for j=1:m_step
49     S=chol(P,'lower');xes_u=xes1(:,j);Alpha1=0;
50     for ji=1:n
51        Ri=P(:,ji); Num=0;Den=0;mu_norm1=0;
52        for jj=1:n
53            mu_norm1=mu_norm1+xes_u(jj)^2;
54            Num=Num+xes_u(jj)*Ri(jj);Den=Den+Ri(jj)^2;
55        end
56        mu_norm=sqrt(mu_norm1); Den1=sqrt(Den);
57        alpha(ji)=((abs(Num))/(Den1*mu_norm));   %% alpha
58        Alpha1=Alpha1+alpha(ji);
59     end
60     m_u=0.8; beta1=max(m_u*alpha); beta2=0.25*beta1-0.5*Alpha1;
61     beta=beta2+.1;
62     Den_pw=Alpha1+beta; m_alpha=m_u*alpha; md_alpha=(1-m_u)*
              alpha;
63     chi(:,1)=xes_u; W(1)=1-((0.5*Alpha1)/Den_pw);
```

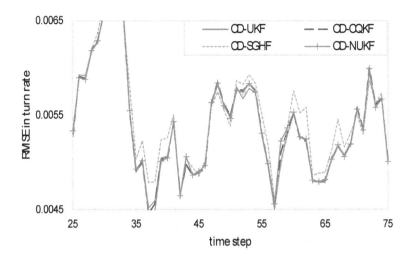

FIGURE 8.10: Problem 3: RMSE of turn rate, Ω.

```
64  % Sigma points and weights calculation
65  for i=1:n
66      chi(:,i+1)=xes_u+sqrt(Den_pw/m_alpha(i))*S(:,i);
67      W(i+1)=(0.25*m_alpha(i))/Den_pw;
68  end
69  for i=n+1:2*n
70      chi(:,i+1)=xes_u-sqrt(Den_pw/m_alpha(i-n))*S(:,i-n);
71      W(i+1)=(0.25*m_alpha(i-n))/Den_pw;
72  end
73  for i=2*n+1:3*n
74      chi(:,i+1)=xes_u+sqrt(Den_pw/md_alpha(i-2*n))*S(:,i-2*n);
75      W(i+1)=(0.25*md_alpha(i-2*n))/Den_pw;
76  end
77  for i=3*n+1:4*n
78      chi(:,i+1)=xes_u-sqrt(Den_pw/md_alpha(i-3*n))*S(:,i-3*n);
79      W(i+1)=(0.25*md_alpha(i-3*n))/Den_pw;
80  end
81  [row col]=size(W);
82  for i=1:col % Process update of sigma points
83      fde_1=chi(1,i)+del*chi(2,i)-del^2/2*chi(7,i)*chi(4,i);
84      fde_2=chi(2,i)-del*chi(7,i)*chi(4,i)-...
85          del^2/2*(chi(7,i)^2*chi(2,i)+(sigma_1*sigma_2/2));
86      fde_3=chi(3,i)+del*chi(4,i)+del^2/2*chi(7,i)*chi(2,i);
87      fde_4=chi(4,i)+del*chi(7,i)*chi(2,i)-...
88          del^2/2*(chi(7,i)^2*chi(4,i)-(sigma_1*sigma_2/2));
89      fde_5=chi(5,i)+del*chi(6,i);fde_6=chi(6,i);fde_7=chi(7,i
            );
90      chinew(:,i)=[fde_1;fde_2;fde_3;fde_4;fde_5;fde_6;fde_7];
91  end
92  meanpred=0;
93  for i=1:col
94      meanpred=meanpred+W(i)*chinew(:,i);
```

```
95    end
96    xes1(:,j+1)=meanpred;
97    L_fe(1,2)=sigma_1;L_fe(2,4)=-sigma_1*xes_u(7);   % L matrix
98    L_fe(2,7)=-sigma_2*xes_u(4);L_fe(3,4)=sigma_1;
99    L_fe(4,2)=sigma_1*xes_u(7);L_fe(4,7)=-sigma_2*xes_u(2);
100   L_fe(5,6)=sigma_1;L_fe(7,7)=0;
101   CHI=0;
102   for i=1:col
103       CHI=CHI+(W(i))*(chinew(:,i)-meanpred)*(chinew(:,i)-
              meanpred)';
104   end
105   P=CHI+del^2/2*(L_fe*sqrt(Q)' + sqrt(Q)*L_fe')+del^3/3*L_fe*
          L_fe'+del*Q;
106   end
107   P_prior=P; meanpred=xes1(:,j+1);
108   S1=chol(P_prior,'lower'); xes_u=meanpred; Alpha1=0;
109   for ji=1:n
110       Ri=P_prior(:,ji); Num=0; Den=0; mu_norm1=0;
111       for jj=1:n
112           mu_norm1=mu_norm1+xes_u(jj)^2;
113           Num=Num+xes_u(jj)*Ri(jj);
114           Den=Den+Ri(jj)^2;
115       end
116       mu_norm=sqrt(mu_norm1); Den1=sqrt(Den);
117       alpha(ji)=((abs(Num))/(Den1*mu_norm));    %% alpha
118       Alpha1=Alpha1+alpha(ji);
119   end
120   m_u=0.8;beta1=max(m_u*alpha);beta2=0.25*beta1-0.5*Alpha1;
121   beta=beta2+.1;Den_pw=Alpha1+beta;m_alpha=m_u*alpha;
122   md_alpha=(1-m_u)*alpha;
123   chii(:,1)=xes_u;Wi(1)=1-((0.5*Alpha1)/Den_pw);
124   % Generation of new sigma points for measurement update
125   for i=1:n
126       chii(:,i+1)=xes_u+sqrt(Den_pw/m_alpha(i))*S1(:,i);
127       Wi(i+1)=(0.25*m_alpha(i))/Den_pw;
128   end
129   for i=n+1:2*n
130       chii(:,i+1)=xes_u-sqrt(Den_pw/m_alpha(i-n))*S1(:,i-n);
131       Wi(i+1)=(0.25*m_alpha(i-n))/Den_pw;
132   end
133   for i=2*n+1:3*n
134     chii(:,i+1)=xes_u+sqrt(Den_pw/md_alpha(i-2*n))*S1(:,i-2*n)
           ;
135     Wi(i+1)=(0.25*md_alpha(i-2*n))/Den_pw;
136   end
137   for i=3*n+1:4*n
138     chii(:,i+1)=xes_u-sqrt(Den_pw/md_alpha(i-3*n))*S1(:,i-3*n);
139     Wi(i+1)=(0.25*md_alpha(i-3*n))/Den_pw;
140   end
141   [row col]=size(Wi);
142   for i=1:col
143       znew(:,i)=[sqrt(chii(1,i)^2+chii(3,i)^2+chii(5,i)^2);atan(
              chii(3,i)...
144           /chii(1,i)); atan(chii(5,i)/sqrt(chii(1,i)^2+chii(3,i)
                 ^2))];
145   end
146   zpred=0;
```

```
147    for i=1:col, zpred=zpred+Wi(i)*znew(:,i);end % Expected
           measurement
148    Pyy=0;
149    for i=1:col, Pyy=Pyy+Wi(i)*(znew(:,i)-zpred)*(znew(:,i)-
           zpred)';end
150    Pyy_new=Pyy+R;Pxy=0;   % Calculation of Pyy
151    for i=1:col
152        Pxy=Pxy+Wi(i)*(chii(:,i)-meanpred)*(znew(:,i)-zpred)';
153    end
154    K=Pxy/(Pyy_new); % Kalman gain
155    xes(:,k+1)=meanpred+K*(y(:,k+1)-zpred);P=P_prior-K*Pyy_new*K
           ';
156    e1(m,k)=((X(1,k)-xes(1,k))^2+(X(3,k)-xes(3,k))^2+(X(5,k)-xes
           (5,k))^2);
157    e2(m,k)=((X(2,k)-xes(2,k))^2+(X(4,k)-xes(4,k))^2+(X(6,k)-xes
           (6,k))^2);
158    e3(m,k)=(X(7,k)-xes(7,k))^2;
159 end % End of time horizon loop
160 pos_error(m)=sqrt((X(1,kmax+1)-xes(1,kmax+1))^2 + ...
161    (X(2,kmax+1)-xes(2,kmax+1))^2+ (X(3,kmax+1)-xes(3,kmax+1))
           ^2);
162 if pos_error(m) < 50
163    TL=TL+1;
164 end
165 end % End of MC loop
166 % Plot truth and estimated states for single run
167 figure(1); plot(X(1,:));hold on;plot(xes(1,:),'r')
168 figure(2); plot(X(2,:));hold on;plot(xes(2,:),'r')
169 figure(3); plot(X(3,:));hold on;plot(xes(3,:),'r')
170 figure(4);plot(X(4,:));hold on;plot(xes(4,:),'r')
171 figure(5);plot(X(5,:));hold on;plot(xes(5,:),'r')
172 figure(6);plot(X(6,:));hold on;plot(xes(6,:),'r')
173 figure(7);plot(X(7,:));hold on;plot(xes(7,:),'r')
174 % Calculation and plot of RMSE
175 tt=(1:kmax)*T;
176 for k=1:kmax,  E=0;
177    for m=1:M, E=E+e1(m,k); end
178    RMSE_pos(k)=(E/M)^0.5;
179 end
180 for k=1:kmax, E=0;
181    for m=1:M, E=E+e2(m,k); end
182    RMSE_vel(k)=(E/M)^0.5;
183 end
184 for k=1:kmax, E=0;
185    for m=1:M, E=E+e3(m,k); end
186    RMSE_omega(k)=(E/M)^0.5;
187 end
188 figure(8);hold on;plot(RMSE_pos,'r')
189 figure(9);hold on;plot(RMSE_vel,'r')
190 figure(10);hold on;plot(RMSE_omega,'r')
```

Listing 8.1: Continuous-discrete filter for 7D target tracking problem.

The performance of various filtering algorithms is evaluated by plotting their RMSEs and by counting the number of diverged tracks. A track is said to be diverged when the position error at last time instant (75th time step) is greater

TABLE 8.3: Problem 3: % of track-loss, number of support points and relative computational time.

Filter	Track-loss (%)	No. of points	Relative comput.time
CD-UKF	4	15	1
CD-CQKF	2	28	1.56
CD-SGHF	6	127	3.72
CD-NUKF	2	29	1.58

than $50m$, i.e., $poserror_{kmax} \geq 50m$. RMSEs of resultant position, velocity and turn rate for 100 Monte Carlo runs have been plotted in Figure 8.8, 8.9 and 8.10 respectively. From these figures, it can be noted that CD-UKF, CD-CQKF and CD-NUKF performed with a comparable accuracy, which is better than CD-SGHF. Accurate tracking performance of CD-NUKF in the continuous-discrete domain is further illustrated by plotting the truth and estimated target trajectory for a time period of $90s$ to $150s$ (see Figure 8.7), where a higher turn rate of $\Omega = 8^o/s$ is considered throughout the study.

The number of diverged tracks for different filters is counted and is mentioned in Table 8.3. From this table, it can be observed that CD-NUKF performs with slightly higher accuracy than CD-UKF, whereas the number of diverged tracks for CD-SGHF is high (6%). The number of support points for the filters under study and the associated computational cost are also mentioned in this table. Since CD-NUKF and CD-CQKF have almost the same number of support points, their computational cost is comparable.

8.5   Summary

The chapter begins with a description of a continuous system where a stochastic differential equation is used to describe the process model and a continuous algebraic equation is used to represent the measurement model. The derivation of the Kalman-Bucy filter is presented for such systems. Then a continuous-discrete system is described where the state evolution is in continuous time whereas the measurements are in discrete time. For a nonlinear continuous-discrete system, main numerical schemes to discretize the process model are discussed. Deterministic sample point filters are described for such systems. Finally, in the example section, three problems are solved with continuous-discrete filters. The MATLAB code of a continuous-discrete filter for a target tracking problem has also been included.

Chapter 9

Case studies

9.1	Introduction	187
9.2	Bearing only underwater target tracking problem	188
9.3	Problem formulation	189
	9.3.1 Tracking scenarios	190
9.4	Shifted Rayleigh filter (SRF)	191
9.5	Gaussian sum shifted Rayleigh filter (GS-SRF)	193
	9.5.1 Bearing density	194
9.6	Continuous-discrete shifted Rayleigh filter (CD-SRF)	194
	9.6.1 Time update of CD-SRF	196
9.7	Simulation results	196
	9.7.1 Filter initialization	199
	9.7.2 Performance criteria	201
	9.7.3 Performance analysis of Gaussian sum filters	201
	9.7.4 Performance analysis of continuous-discrete filters	211
9.8	Summary	215
9.9	Tracking of a ballistic target	216
9.10	Problem formulation	219
	9.10.1 Process model	219
	9.10.1.1 Process model in discrete domain	219
	9.10.1.2 Process model in continuous time domain	220
	9.10.2 Seeker measurement model	220
	9.10.3 Target acceleration model	223
9.11	Proportional navigation guidance (PNG) law	225
9.12	Simulation results	226
	9.12.1 Performance of adaptive Gaussian sum filters	228
	9.12.2 Performance of continuous-discrete filters	229
9.13	Conclusions	230

9.1 Introduction

In the previous chapters, we have discussed various Gaussian filters which are available in recent literature. The filters are applied to different sample problems or simplified problems which have little resemblance to a real scenario. In this chapter, we consider the following two real life problems:

(i) Underwater target motion analysis with the help of a passive angle only sensor. The problem is popularly known as bearing only tracking (BOT).

(ii) Tracking and interception of a ballistic target in the terminal guidance phase with the help of seeker measurements.

Almost all the deterministic sample point filters are applied on the two problems and their performance is compared here.

9.2   Bearing only underwater target tracking problem

Bearings-only tracking (BOT) is one of the most challenging nonlinear filtering problems. The main advantage of such type of tracking is that the observer's location is not exposed to the vehicle which is being tracked because bearing angle measurement is a passive measurement. BOT plays a very important role particularly in tracking enemy ships and submarines, etc. Here, we shall formulate a problem where the enemy ship will be tracked from own ship. The objective is to find the range (positions) and velocity of the enemy ship. This problem is often termed target motion analysis (TMA). If a single observer is used to track it is called autonomous TMA [9]. For single observer BOT, the observer has to maneuver in order to make the system observable [150, 45, 75].

The target is modeled by a general state-space representation that consists of a system model for the dynamics and an observation model for the measurement. Mainly two different target models are considered for BOT problems which are the constant velocity (CV) and coordinated turn (CT) models [130]. As the name suggests, in the CV model, target velocity is constant and in the CT model the target takes the turn with nearly constant angular velocity. Most of the work has been done with the CV model which we consider here.

Since the measurement is only the noise corrupted bearing angle, the problem is highly nonlinear and challenging. Almost all the nonlinear filters have been applied to the underwater BOT problem [130, 100, 124, 114] and their performances have been compared.

A new filter named the shifted Rayleigh filter (SRF) [29, 30] has been developed exclusively for the BOT problem where an exact calculation of the posterior mean and covariance can be obtained. SRF is the best filter for BOT problems. In this chapter, we shall discuss the SRF algorithm. SRF will further be modified and combined with the Gaussian sum filter. We call it the Gaussian sum shifted Rayleigh filter (GS-SRF). One more modification of SRF is made here so that it can be applied to a BOT with a continuous process model. This is called the continuous-discrete shifted Rayleigh filter (CD-SRF).

In this chapter, an autonomous TMA problem is considered. Apart from the particle filter, almost all the filters have been implemented (including SRF,

GS SRF). The performance of these filters was studied on two target-observer scenarios, and their performance was compared with other suboptimal solutions. The performance comparison is done by calculating the RMSE, track-loss and computational time. Further, a parametric study was carried out by varying the initial covariance, measurement noise covariance and sampling time to validate the filtering accuracy.

9.3 Problem formulation

This section describes the modeling of an autonomous TMA problem where a constant velocity target has to be tracked using bearing measurements which are noise corrupted. Here, the measurements are obtained from a single moving ship, referred to as the observer. The problem is mathematically defined in the Cartesian coordinate system, where the motive is to estimate the position and velocity of the target, defined as $\mathcal{X}^t = [x^t \ y^t \ v^t_x \ v^t_y]^T$. Similarly, observer state variables are defined as $\mathcal{X}^o = [x^o \ y^o \ v^o_x \ v^o_y]^T$. Now, a relative state vector can be defined as $\mathcal{X}_k \triangleq \mathcal{X}^t_k - \mathcal{X}^o_k$.

The process model in the continuous domain can be represented as $\dot{\mathcal{X}} = F(\mathcal{X})$, where $F(\mathcal{X}) = [v_x \ v_y \ 0 \ 0]^T$ and (v_x, v_y) are relative velocity vectors in x and y directions, respectively. Performing a simple discretization procedure on the continuous model and adding white Gaussian noise, the discretized model becomes

$$\mathcal{X}_{k+1} = \mathcal{F}\mathcal{X}_k + \eta_k - \mho_{k,k+1}, \tag{9.1}$$

where $\mathcal{F}$ is the state transition matrix given as

$$\mathcal{F} = \begin{bmatrix} I_{2\times2} & TI_{2\times2} \\ O_{2\times2} & I_{2\times2} \end{bmatrix}, \tag{9.2}$$

and the covariance matrix Q of process noise η_k is

$$Q = \tilde{q} \begin{bmatrix} \frac{T^3}{3}I_{2\times2} & \frac{T^2}{2}I_{2\times2} \\ \frac{T^2}{2}I_{2\times2} & TI_{2\times2} \end{bmatrix}, \tag{9.3}$$

where T is the sampling interval, $O_{2\times2}$ is a zero matrix of order 2 and $\tilde{q}$ is the process noise intensity. To account for the observer manoeuvre in the target dynamics, $\mho_{k,k+1}$ is defined which is a vector of observer inputs, given by

$$\mho_{k,k+1} = \begin{bmatrix} x^o_{k+1} - x^o_k - Tv^o_{x,k} \\ y^o_{k+1} - y^o_k - Tv^o_{y,k} \\ v^o_{x,k+1} - v^o_{x,k} \\ v^o_{y,k+1} - v^o_{y,k} \end{bmatrix}. \tag{9.4}$$

The observer locations are usually obtained by an on-board inertial navigation system powered by a global positioning system (GPS). Now the measurement equation can be expressed as

$$\mathcal{Y}_k = \gamma(\mathcal{X}_k) + v_k, \tag{9.5}$$

where v_k is modeled as a Gaussian noise with zero mean and standard deviation σ_θ. From this model, the true bearing measurements, $\gamma(\mathcal{X}_k) = \theta_k = \tan^{-1}\left(\frac{x_k}{y_k}\right)$ are obtained. Traditionally the speed in the underwater scenario is expressed in knots, one of which is equivalent to 1.85 Km/hr. The course of the vehicle means the angle of the displacement vector from the north direction. Bearing angle is also measured with the reference of north direction.

9.3.1   Tracking scenarios

Two tracking scenarios have been considered as shown in Figures 9.1 and 9.2. Both the scenarios are taken from open literature [100]. In the first engagement scenario, a moderately nonlinear measurement model is considered. The second scenario becomes highly nonlinear because we consider an abrupt change in the bearing, which makes it more challenging for the filters to track the target. The values assumed for all the parameters in simulating the two target-observer scenarios are tabulated in Table 9.1.

TABLE 9.1: Tracking scenario parameters [124]

Parameters	Case 1	Case 2
Initial range (r)	5 km	10 km
Target speed $(\bar{s})$	4 $knots$	15 $knots$
Target course	-140°	-135.4^o
Observer speed	5 $knots$	5 $knots$
Observer initial course	140°	-80°
Observer final course	20°	146°
Observer manoeuvre	From 13th to 17th min	15th min
σ_θ	1.5^o	2°
σ_c	$\pi/\sqrt{12}$	$\pi/\sqrt{12}$
$\tilde{q}$	$1.944 \times 10^{-6} km^2/min^3$	$2.142 \times 10^{-6} km^2/min^3$
σ_r	2 km	4 km
σ_s	2 $knots$	2 $knots$

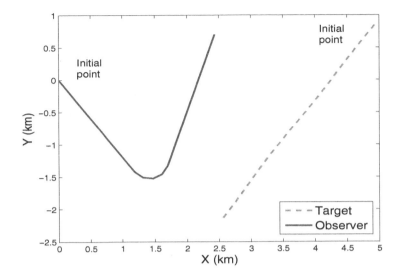

FIGURE 9.1: Target and observer dynamics for the first scenario [124].

9.4 Shifted Rayleigh filter (SRF)

The shifted Rayleigh filter has proved to be a highly accurate and computationally efficient nonlinear filter for solving BOT problems. This is achieved by reformulating the conventional measurement equation in terms of an augmented measurement, which is the relative position vector between the target and the observer. Given a Gaussian prior, it finds out the exact mean and covariance of the posterior density. At the end of each Kalman filtering cycle, this exact mean and covariance are used to approximate the non-Gaussian posterior as a Gaussian density.

Apart from the process and measurement models described earlier, SRF defines an augmented measurement model, which is the relative position vector between target and observer. This can be represented as $\bar{\mathcal{Y}}_k = H\mathcal{X}_k + \bar{v}_k$, where $\bar{v}_k$ is a Gaussian process noise vector with zero mean and covariance matrix $R_k = \sigma_\theta^2 \mathbb{E}[\| H\mathcal{X}_k \|^2] I_{2\times 2}$, and $H = [I_{2\times 2} \ O_{2\times 2}]$. Here, σ_θ is the standard deviation of white noise corrupting the true measurement, θ_k. Now consider a transformed measurement, called the actual measurement; $\mathbf{b}_k$ are a vector of direction cosines to the state $\mathcal{X}_k$ corrupted with noise. It can also be interpreted as the projection of $\bar{\mathcal{Y}}_k$ on to the unit circle, that is $\mathbf{b}_k = \dfrac{\bar{\mathcal{Y}}_k}{\| \bar{\mathcal{Y}}_k \|} =$ $[\sin(\mathcal{Y}_k) \ \cos(\mathcal{Y}_k)]^T$, where $\| \bar{\mathcal{Y}}_k \|$ is the range. SRF equations are derived assuming direct access to the augmented measurement $\bar{\mathcal{Y}}_k$. However, instead

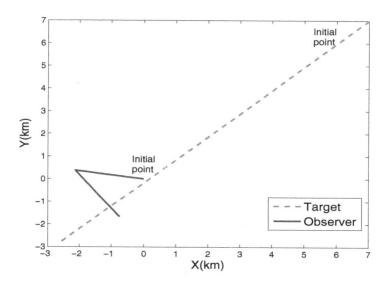

FIGURE 9.2: Target and observer dynamics for the second scenario. [125]

of $\bar{\mathcal{Y}}_k$, b_k is the only available measurement. Hence, filtering equations are modified so that the unavailability of $\bar{\mathcal{Y}}_k$ is accounted for by evaluating the moments of $\mathcal{X}_k$ conditioned on b_k. Here, the author chooses to write the prior and posterior densities of $\mathcal{X}_k$ conditioned on $\bar{\mathcal{Y}}_k$.

Now the prior density given can be represented as Eq. (9.6). To find out this density, the mean $\hat{\mathcal{X}}_{k+1|k}$ and the covariance $\Sigma_{k+1|k}$ have to be calculated. These are obtained from the SRF equations as

$$\hat{\mathcal{X}}_{k+1|k} = \mathcal{F}\hat{\mathcal{X}}_{k|k} - \mathcal{U}_{k,k+1}$$

and

$$\Sigma_{k+1|k} = \mathcal{F}\Sigma_{k|k}\mathcal{F}^T + Q.$$

Similarly, the posterior density for GS-SRF formulation is expressed as Eq. (9.7). The posterior mean and covariance are defined as

$$\hat{\mathcal{X}}_{k+1|k+1} = \hat{\mathcal{X}}_{k+1|k} + K_k[\zeta_{k+1}b_{k+1} - H\hat{\mathcal{X}}_{k+1|k}]$$

and

$$\Sigma_{k+1|k+1} = (\mathbf{I} - K_k H)\Sigma_{k+1|k} + \Psi_{k+1}K_k b_{k+1}b_{k+1}^T K_k^T,$$

where the Kalman gain $K_k = \Sigma_{k+1|k}H^T V_{k+1}^{-1}$ and $V_{k+1} = H\Sigma_{k+1|k}H^T + R_k$. Here,

$$\zeta_{k+1} = \mathbb{E}[\| \bar{\mathcal{Y}}_{k+1} \| |b_{k+1}] = [b_{k+1}^T V_{k+1}^{-1} b_{k+1}]^{-1/2}\rho(\mathsf{z}_{k+1})$$

and

$$\Psi_{k+1} = \text{var}[\| \bar{\mathcal{Y}}_{k+1} \| |b_{k+1}] = [b_{k+1}^T V_{k+1}^{-1} b_{k+1}]^{-1}(2 + \mathsf{z}_{k+1}\rho(\mathsf{z}_{k+1}) - \rho(\mathsf{z}_{k+1})^2),$$

where the variable $z_{k+1} = [b_{k+1}^T V_{k+1}^{-1} b_{k+1}]^{-1/2} b_{k+1}^T V_{k+1}^{-1} [H \hat{\mathcal{X}}_{k+1|k}]$. In these equations, $\rho(z_{k+1})$ is the mean of a shifted Rayleigh variable [29] expressed as

$$\rho(z_{k+1}) = \frac{\int_0^\infty s^2 \exp\left\{\frac{-1}{2}(s - z_{k+1})^2\right\} ds}{\int_0^\infty s \exp\left\{\frac{-1}{2}(s - z_{k+1})^2\right\} ds}$$

$$= \frac{z_{k+1} \exp\left\{-z_{k+1}^2/2\right\} + \sqrt{2\pi}(z_{k+1}^2 + 1)\mathbb{F}(z_{k+1})}{\exp\left\{-z_{k+1}^2/2\right\} + \sqrt{2\pi} z_{k+1} \mathbb{F}(z_{k+1})},$$

where $\mathbb{F}(z_{k+1})$ is the cumulative distribution function of a standard normal variable.

9.5   Gaussian sum shifted Rayleigh filter (GS-SRF)

As we discuss in Chapter 6, in the Gaussian sum approach, prior and posterior pdfs are represented as the weighted sum of individual Gaussian pdf [100, 95]. The prior pdf could be expressed as

$$p(\mathcal{X}_{k+1}|\mathcal{Y}_k) \approx \sum_{i=1}^{N_c} \bar{w}_{k+1|k}^i \mathcal{N}(\mathcal{X}_{k+1}; \hat{\mathcal{X}}_{k+1|k}^i, \Sigma_{k+1|k}^i), \qquad (9.6)$$

where N_c is the number of Gaussian components, $\bar{w}^i$ is their corresponding weights and $i = 1, \cdots, N_c$. The individual Gaussian densities $\mathcal{N}(\mathcal{X}_{k+1}; \hat{\mathcal{X}}_{k+1|k}^i, \Sigma_{k+1|k}^i)$ are generated with the help of SRF.

The posterior pdf can be written as

$$p(\mathcal{X}_{k+1}|\mathcal{Y}_{k+1}) \propto p(\mathcal{Y}_{k+1}|\mathcal{X}_{k+1}) p(\mathcal{X}_{k+1}|\mathcal{Y}_k)$$

$$\approx \sum_{i=1}^{N_c} \bar{w}_{k+1|k+1}^i \mathcal{N}(\mathcal{X}_{k+1}; \hat{\mathcal{X}}_{k+1|k+1}^i, \Sigma_{k+1|k+1}^i), \qquad (9.7)$$

where the update of weights after measurement is received, is done using

$$\bar{w}_{k+1|k+1}^i = \frac{p(\mathcal{Y}_{k+1}|\mathcal{X}_{k+1}, i) \bar{w}_{k|k}^i}{\sum_{i=1}^{N_c} p(\mathcal{Y}_{k+1}|\mathcal{X}_{k+1}, i) \bar{w}_{k|k}^i}. \qquad (9.8)$$

The posterior means $\hat{\mathcal{X}}_{k+1|k+1}^i$ and covariances $\Sigma_{k+1|k+1}^i$ of the ith Gaussian component can be determined with the help SRF. In Chapter 6, we have seen that in the ordinary GS filter, the weights of individual Gaussian components are updated only during the measurement update step. To update the weights of the Gaussian components in the presence of measurement, the relation given above has to be calculated for which determination of bearing density is required.

9.5.1   Bearing density

In this section, the bearing density which is essential for the formulation of GS-SRF is derived. Let $\bar{\mathcal{Y}}_{k+1} = [x_{k+1} \; y_{k+1}]^T$ be a Gaussian random variable with mean $m = [m_1 \; m_2]^T = H\hat{\mathcal{X}}_{k+1|k}$ and covariance V. Then its distribution in polar coordinates can be expressed as

$$p(r,\theta) = \frac{r}{2\pi\sqrt{|V|}} \exp\left\{ -1/2\left(g(r,\theta) - m\right)^T V^{-1}\left(g(r,\theta) - m\right)\right\},$$

where $g(r,\theta) = [r\cos(\theta) \; r\sin(\theta)]^T$. This can be simplified as

$$p(r,\theta) = \frac{r}{2\pi\sqrt{|V|}} \exp\left\{ -1/2(\mathsf{A}r^2 + \mathsf{B}r + \mathsf{C})\right\},$$

where

$\mathsf{A} = a_{11}\sin^2(\theta) + (a_{12} + a_{21})\cos(\theta)\sin(\theta) + a_{22}\cos^2(\theta),$

$\mathsf{B} = -2a_{11}m_1\sin(\theta) - 2a_{22}m_2\cos(\theta) - (a_{12} + a_{21})(m_1\cos(\theta) + m_2\sin(\theta)),$

$\mathsf{C} = a_{11}m_1^2 + a_{21}m_1m_2 + a_{12}m_1m_2 + a_{22}m_2^2,$

and a_{ij} are the elements of V^{-1}.

Rearranging $p(r,\theta)$ we get

$$p(r,\theta) = \frac{r}{2\pi\sqrt{|V|}} \exp\left\{ \frac{\mathsf{B}^2}{8\mathsf{A}} - \frac{\mathsf{C}}{2}\right\} \exp\left\{ -\frac{\mathsf{A}}{2}\left(r + \frac{\mathsf{B}}{2\mathsf{A}}\right)^2\right\}.$$

The density of θ can be found by integrating with respect to r,

$$p(\theta) = \frac{1}{2\pi\sqrt{|V|}} \exp\left\{ \frac{\mathsf{B}^2}{8\mathsf{A}} - \frac{\mathsf{C}}{2}\right\} \int_0^\infty r\exp\left\{ -\frac{\mathsf{A}}{2}\left(r + \frac{\mathsf{B}}{2\mathsf{A}}\right)^2\right\}dr$$

$$= \frac{\exp\left\{ \frac{\mathsf{B}^2}{8\mathsf{A}} - \frac{\mathsf{C}}{2}\right\}}{2\pi\sqrt{|V|}} \left(\frac{1}{\mathsf{A}}\exp\left\{ -\frac{\mathsf{B}^2}{8\mathsf{A}}\right\} - \frac{\sqrt{2\pi}\mathsf{B}}{2\mathsf{A}^{3/2}}\mathbb{F}\left(-\frac{\mathsf{B}}{2\sqrt{\mathsf{A}}}\right)\right).$$

The algorithm of GS-SRF is quite straightforward and is presented with the help of a flowchart illustrated in Figure 9.3. For more details, interested readers are requested to read [124].

9.6   Continuous-discrete shifted Rayleigh filter (CD-SRF)

In this section, SRF is formulated to deal with BOT problems where the process is represented by a stochastic differential equation (in continuous domain) and the measurement is discrete [125]. The importance of continuous-discrete filtering and its advantages are mentioned in Chapter 8. Consider the

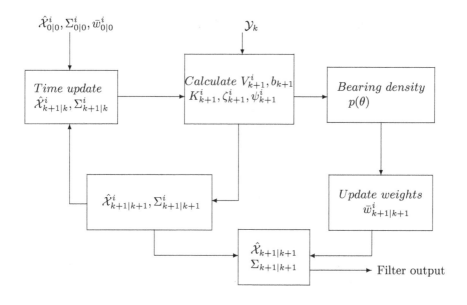

FIGURE 9.3: Flow chart for GS-SRF algorithm.

continuous process function $f(\mathcal{X}) = [v_x \; v_y \; 0 \; 0]^T$. The corresponding stochastic differential equation is expressed as Eq. (8.1). For the successful implementation of the filtering algorithms, the continuous process model has to be discretized effectively. Here, the Ito-Taylor expansion of order 1.5 is used.

Applying the Ito-Taylor expansion of order 1.5 explained in Section 8.3.3 for the time interval $(t, t+\delta)$, then $f_d(\mathcal{X}) = \mathcal{F}\mathcal{X}$ (see Eq. (9.2) for the expression of $\mathcal{F}$), $\mathbb{L}_0 = 0$ and

$$\mathbb{L} = \sqrt{\tilde{q}} \left[\begin{array}{cc} \sqrt{\frac{T^2}{2}} \mathrm{I}_{2\times 2} & \sqrt{T} \mathrm{I}_{2\times 2} \\ 0_{2\times 2} & 0_{2\times 2} \end{array} \right].$$

Now the resultant discretized model can be derived as

$$\mathcal{X}_k^{j+1} = f_d(\mathcal{X}_k^j) + \sqrt{Q}\eta + \mathbb{L}\eta' - \mho_{k,k+1}, \qquad (9.9)$$

where $\mathcal{X}_k^j$ denotes $\mathcal{X}(t)$ at time $t = t_k + j\delta$, with $1 \le j \le m_{step}$, $\delta = T/m_{step}$ and $k = 1, \cdots, k_{max}$. The expressions for Q and $\mho_{k,k+1}$ are mentioned in Eq. (9.3) and Eq. (9.4) respectively.

9.6.1 Time update of CD-SRF

Assuming that the statistics of $\mathcal{X}_k$ follows a normal distribution with mean $\hat{\mathcal{X}}_{k|k}$ and covariance $\Sigma_{k|k}$, and considering the process model given in Eq. (9.9), the predicted state estimate can be expressed as

$$\hat{\mathcal{X}}_{k|k}^{j+1} = \mathbb{E}[f_d(\mathcal{X}_k^j) + \sqrt{Q}\eta + \mathbb{L}\eta' - \mho_{k,k+1}]$$

$$= \mathcal{F}\hat{\mathcal{X}}_{k|k}^j - \mho_{k,k+1}.$$

Similarly, the predicted state error covariance matrix can be expressed as

$$\Sigma_{k|k}^{j+1} = \mathbb{E}[(\mathcal{X}_{k|k}^{j+1} - \hat{\mathcal{X}}_{k|k}^{j+1})(\mathcal{X}_{k|k}^{j+1} - \hat{\mathcal{X}}_{k|k}^{j+1})^T]$$

$$= \mathbb{E}[(\mathcal{F}(\mathcal{X}_k^j - \hat{\mathcal{X}}_{k|k}^j) + \sqrt{Q}\eta + \mathbb{L}\eta')(\mathcal{F}(\mathcal{X}_k^j - \hat{\mathcal{X}}_{k|k}^j) + \sqrt{Q}\eta + \mathbb{L}\eta')^T]$$

$$= \mathcal{F}\Sigma_{k|k}^j \mathcal{F}^T + \delta Q + \frac{\delta^2}{2}[\sqrt{Q}\mathbb{L}^T + \mathbb{L}\sqrt{Q}^T] + \frac{\delta^3}{3}\mathbb{L}\mathbb{L}^T.$$

Then the prior mean and covariance can be obtained as $\hat{\mathcal{X}}_{k+1|k} = \hat{\mathcal{X}}_{k|k}^{m_{step}}$ and $\Sigma_{k+1|k} = \Sigma_{k|k}^{m_{step}}$. An algorithm for CD-SRF is provided in Algorithm 13.

9.7 Simulation results

The filtering performance of various filters including SRF, GS-SRF and CD-SRF is studied by implementing them on two tracking scenarios described in Section 9.3.1. Further, the continuous-discrete sparse-grid Gauss-Hermite (CD-SGHF) and continuous-discrete new unscented Kalman filter (CD-NUKF) [128] are also implemented. A typical MATLAB code for CKF in BOT is provided in Listing 9.1.

```
1 % Estimation of range and velocity of a BOT problem
2 % Scenario 1 from Arulampalam, A. Lamahewa, T.D. Abhayapala,
3 %'A Gaussian-Sum Based Cubature Kalman Filter for Bearings-Only
      Tracking",
4 % IEEE TRANSACTIONS ON AEROSPACE AND ELECTRONIC SYSTEMS VOL. 49,
5 % NO. 2 APRIL 2013
6 % Program for Monte Carlo run
7 % Code credit: Rahul Radhakrishnan,IIT Patna
8 %
9 clear all;
10 %randn('seed',1) Seeding random number generator if required
11 MC=100;   % Number of Monte Carlo run
12 kmax=30;  % Maximum iteration
13 TLL=0;    % Required to calculate Track-loss count
14 T=1;      % Sampling time
15 q=[T^3/3 0 T^2/2 0;0 T^3/3 0 T^2/2;T^2/2 0 T 0;0 T^2/2 0 T];
16 Q=(2.944*10^-6)*q; % Process noise covariance
17 sigmatheta=1.5*pi/180;  %Measurement noise covariance
18 F =[1 0 T 0;0 1 0 T;0 0 1 0;0 0 0 1]; % System    matrix
```

```
19% Generate observer trajectory ——————————————————
20xt(:,1)=[4.9286;.8420;(0.123466*sin(-140*pi/180));...
21    (0.123466*cos(-140*pi/180))]; % Target state initialize
22xo(:,1)=[0;0;0.15433*sin(140*pi/180);0.15433*cos(140*pi/180)];
23for k=1:12
24    xo(1,k+1)=xo(1,k)+0.15433*sin(140*pi/180)*T;
25    xo(2,k+1)=xo(2,k)+0.15433*cos(140*pi/180)*T;
26    xo(3,k+1)=xo(3,k);
27    xo(4,k+1)=xo(4,k);
28end
29for k=13
30    xo(1,k+1)=xo(1,k)+0.15433*sin(125*pi/180)*T;
31    xo(2,k+1)=xo(2,k)+0.15433*cos(125*pi/180)*T;
32    xo(3,k+1)=0.15433*sin(125*pi/180);
33    xo(4,k+1)=0.15433*cos(125*pi/180);
34end
35for k=14
36    xo(1,k+1)=xo(1,k)+0.15433*sin(96*pi/180)*T;
37    xo(2,k+1)=xo(2,k)+0.15433*cos(96*pi/180)*T;
38    xo(3,k+1)=0.15433*sin(96*pi/180);
39    xo(4,k+1)=0.15433*cos(96*pi/180);
40end
41for k=15
42    xo(1,k+1)=xo(1,k)+0.15433*sin(65*pi/180)*T;
43    xo(2,k+1)=xo(2,k)+0.15433*cos(65*pi/180)*T;
44    xo(3,k+1)=0.15433*sin(65*pi/180);
45    xo(4,k+1)=0.15433*cos(65*pi/180);
46end
47for k=16
48    xo(1,k+1)=xo(1,k)+0.15433*sin(34*pi/180)*T;
49    xo(2,k+1)=xo(2,k)+0.15433*cos(34*pi/180)*T;
50    xo(3,k+1)=0.15433*sin(34*pi/180);
51    xo(4,k+1)=0.15433*cos(34*pi/180);
52end
53for k=17:kmax
54    xo(1,k+1)=xo(1,k)+0.15433*sin(20*pi/180)*T;
55    xo(2,k+1)=xo(2,k)+0.15433*cos(20*pi/180)*T;
56    xo(3,k+1)=0.15433*sin(20*pi/180);
57    xo(4,k+1)=0.15433*cos(20*pi/180);
58end
59%———— Generation of truth states and measurements ————
60for j=1:MC % Starting of Monte Carlo loop
61x(:,1)=xt(:,1)-xo(:,1);   % Relative state vector
62xtrue(j).relst(:,1)=x(:,1); % Structure to store relative and
        true state
63xtrue(j).tarst(:,1)=xt;
64y(j,1)=(atan2((x(1,1)),(x(2,1))))+0+sigmatheta*randn(1); %
        Initial measurement
65for k=1:kmax
66  w1=0+sigmatheta*randn(1);
67x(:,k+1)=F*x(:,k)+(mvnrnd(zeros(4,1),Q))'-[xo(1,k+1)-xo(1,k)-...
68(T*xo(3,k));xo(2,k+1)-xo(2,k)-(T*xo(4,k));xo(3,k+1)-xo(3,k);...
69        xo(4,k+1)-xo(4,k)];
70    xtrue(j).relst(:,k+1)=x(:,k+1); % Relative truth state
71    xtrue(j).tarst(:,k+1)=x(:,k+1)+xo(:,k+1);
72%     xtrue(j,:,k)=x(:,k+1);
73y(j,k+1)=(atan2((x(1,k+1)),(x(2,k+1))))+w1; %Bearing measurement
```

```
74 end
75 %xt=x+xo;
76    end  % End of truth Monte Truth Carlo loop
77 % Plot a typical engagement scenario
78 figure(1)
79 plot(xo(1,:),xo(2,:),'b—');hold on
80 plot(xtrue(j).tarst(1,:),xtrue(j).tarst(2,:),'r—')
81
82 %————————— Estimation loop ——————————
83 for j=1:MC % Start of estimation Monte Carlo loop
84 r=5; sigmar=2; rr=r+sigmar*randn(1); theta=y(j,1);
85 theta0=theta+sigmatheta*randn(1); s=0.12346;
86 sigmas=.061732; ss=s+sigmas*randn(1); cc=theta0+pi;
87 sigmac=pi/sqrt(12);
88 xes(:,1)=[rr*sin(theta0);rr*cos(theta0);ss*sin(cc)-(xo(3,1))
         ;...
89       ss*cos(cc)-(xo(4,1))]; % Initial estimate
90 Px=(rr^2*sigmatheta^2*(cos(theta0))^2)+(sigmar^2*(sin(theta0))
         ^2);
91 Py=(rr^2*sigmatheta^2*(sin(theta0))^2)+(sigmar^2*(cos(theta0))
         ^2);
92 Pxy=(sigmar^2-(rr^2*sigmatheta^2))*(sin(theta0))*(cos(theta0));
93 Pxx=(ss^2*sigmac^2*(cos(cc))^2)+(sigmas^2*(sin(cc))^2);
94 Pyy=(ss^2*sigmac^2*(sin(cc))^2)+(sigmas^2*(cos(cc))^2);
95 Pxyx=(sigmas^2-(ss^2*sigmac^2))*cos(cc)*sin(cc);
96 P=[Px Pxy 0 0;Pxy Py 0 0;0 0 Pxx Pxyx;0 0 Pxyx Pyy]; % Initial
         error cov
97
98 n=4;[cp, W]=cubquad_points(n,1);cp=cp'; %% points, wts with CKF
         [np, xx]=size(W);
99 E(:,1)=x(:,1)-xes(:,1);
100 normal_err(j,1)=E(:,1)'*inv(P)*E(:,1);
101 xtt(:,1)=xes(:,1)+xo(:,1);
102 for k=1:kmax % Start of estimation algorithm time loop
103 meanpred=F*xes(:,k)-[xo(1,k+1)-xo(1,k)-(T*xo(3,k));...
104      xo(2,k+1)-xo(2,k)-(T*xo(4,k));xo(3,k+1)-xo(3,k);xo(4,k+1)-xo
         (4,k)];
105 Pnew1=F*P*F'+Q;
106 p1=chol(Pnew1,'lower'); % Cholesky factorization
107 for i=1:np chii(:,i)=meanpred+p1*cp(:,i); end
108 for i=1:np znew1(i)=atan2((chii(1,i)),(chii(2,i))); end
109 zpred=0;
110 for i=1:np zpred=zpred+W(i)*znew1(i); end
111 Pnewm=0;
112  for i=1:np Pnewm=Pnewm+(W(i)*(znew1(i)-zpred)*(znew1(i)-zpred)
         '); end
113 Pnewm1=Pnewm+sigmatheta^2; Pxznew=0;
114  for i=1:np % Calculation of PXY
115·     Pxznew=Pxznew+(W(i)*(chii(:,i)-meanpred)*(znew1(i)-zpred)')
         ;
116 end
117 K=Pxznew/Pnewm1; %Kalman gain
118 xes(:,k+1)=meanpred+(K*(y(j,k+1)-zpred)); %Posterior estimate
119 P=Pnew1-(K*Pnewm1*K'); % Posterior error cov
120 xtt(:,k+1)=xes(:,k+1)+xo(:,k+1); % Calculate target state
121
122 end % End of time step loop
```

```
123 e=xtt-xtrue(j).tarst; % Calculation of error matrix after a run
124 Error_sqr=e.^2;  % Square of error matrix
125 esum(j,:)=Error_sqr(1,:)+Error_sqr(2,:);
126 poserror(j,:)=sqrt(esum(j,:));  % Range of error
127
128 % Exclude Track-loss
129 end_range(j)=sqrt((xtt(1,kmax+1)-xtrue(j).tarst(1,kmax+1))^2
        ...
130     + (xtt(2,kmax+1)-xtrue(j).tarst(2,kmax+1))^2);
131 if end_range(j) < 1
132     TLL=TLL+1
133     eetr(TLL,:)=esum(j,:);
134 end
135
136 end  %End of Monte Carlo loop
137
138 for k=1:kmax % Calculation of RMSE
139     e2=0;
140     for j=1:TLL e2=e2+eetr(j,k); end
141     e3(k)=sqrt((e2/TLL));  % RMSE position
142 end
143 figure(2); plot(e3);  % RMSE plot
144 % ——————————————END of the CODE ——————————————
```

Listing 9.1: Generation of RMSE for a bearing only tracking problem.

The simulation results and its discussions are made on the performance of Gaussian sum filters and continuous-discrete filters. Later, a conclusion regarding the performance of all these filters is derived. The filtering performance of all the filters is studied and validated over 10000 Monte Carlo runs. For each run, the truth target states are generated in the computer and the filters are implemented with MATLAB software.

9.7.1 Filter initialization

All the filters are initialized using the method given in [100]. The position estimates of the relative state vector are initialized based on the first bearing measurement θ_0 and initial range r. Depending on how the target is closing in on the observer, the initial course estimate can be defined as $\bar{c} = \theta_0 + \pi$ and the initial target speed as $\bar{s}$. To accommodate the ambiguity in speed and range assumption of the target, σ_s and σ_r are considered. The relative state vector can be initialized as

$$\hat{\mathcal{X}}_{0|0} = \begin{bmatrix} \hat{x} \\ \hat{y} \\ \hat{v}_x \\ \hat{v}_y \end{bmatrix} = \begin{bmatrix} r\sin(\theta_0) \\ r\cos(\theta_0) \\ \bar{s}\sin(\bar{c}) - v^o_{x,0} \\ \bar{s}\cos(\bar{c}) - v^o_{y,0} \end{bmatrix} \tag{9.10}$$

and the initial covariance is defined as

$$\Sigma_{0|0} = \begin{bmatrix} \Sigma_{xx} & \Sigma_{xy} & 0 & 0 \\ \Sigma_{yx} & \Sigma_{yy} & 0 & 0 \\ 0 & 0 & \Sigma_{v_x v_x} & \Sigma_{v_x v_y} \\ 0 & 0 & \Sigma_{v_x v_y} & \Sigma_{v_y v_y} \end{bmatrix}, \tag{9.11}$$

Algorithm 13 Continuous discrete shifted Rayleigh filter

- Initialize with $\hat{\mathcal{X}}_{0|0}$ and $\Sigma_{0|0}$.

- For $k = 1 : k_{max}$

  Time update

  - $\hat{\mathcal{X}}^1_{k|k} = \hat{\mathcal{X}}_{k|k}$ and $\Sigma^1_{k|k} = s_{k|k}$

- For $j =: m_{step}$

  - $\hat{\mathcal{X}}^{j+1}_{k|k} = \mathcal{F}\hat{\mathcal{X}}^j_{k|k}$

  - $\Sigma^{j+1}_{k|k} = \mathcal{F}\Sigma^j_{k|k}\mathcal{F}^T + \delta Q + \frac{\delta^2}{2}[\sqrt{Q}\mathbb{L}^T + \mathbb{L}\sqrt{Q}^T] + \frac{\delta^3}{3}\mathbb{L}\mathbb{L}^T$

- End For

  - $\hat{\mathcal{X}}_{k+1|k} = \hat{\mathcal{X}}^{j+1}_{k|k} - \mho_{k,k+1}$

  - $\Sigma_{k+1|k} = \Sigma^{j+1}_{k|k}$

  Measurement update

  - $V_{k+1} = H\Sigma_{k+1|k}H^T + R_k$

  - Determine K_k, $\rho(\mathbf{z}_{k+1})$, $\mathbf{z}_{k+1}$, b_{k+1} as described in section 9.4

  - $\zeta_{k+1} = [b^T_{k+1}V^{-1}_{k+1}b_{k+1}]^{-1/2}\rho(\mathbf{z}_{k+1})$

  - $\Psi_{k+1} = [b^T_{k+1}V^{-1}_{k+1}b_{k+1}]^{-1}(2 + \mathbf{z}_{k+1}\rho(\mathbf{z}_{k+1}) - \rho(\mathbf{z}_{k+1})^2)$

  - $\hat{\mathcal{X}}_{k+1|k+1} = \hat{\mathcal{X}}_{k+1|k} + K_k[\zeta_{k+1}b_{k+1} - H\hat{\mathcal{X}}_{k+1|k}]$

  - $\Sigma_{k+1|k+1} = (\mathbb{I} - K_kH)\Sigma_{k+1|k} + \psi_{k+1}K_kb_{k+1}b^T_{k+1}K^T_k$

- End for

---

where $\Sigma_{xx} = r^2\sigma_\theta^2 \cos^2(\theta_0) + \sigma_r^2 \sin^2(\theta_0)$, $\Sigma_{yy} = r^2\sigma_\theta^2 \sin^2(\theta_0) + \sigma_r^2 \cos^2(\theta_0)$, $\Sigma_{xy} = \Sigma_{yx} = (\sigma_r^2 - r^2\sigma_\theta^2) \sin\theta_0 \cos\theta_0$, $\Sigma_{v_x v_x} = \bar{s}^2\sigma_c^2 \cos^2\bar{c} + \sigma_s^2 \sin^2\bar{c}$, $\Sigma_{v_y v_y} = \bar{s}^2\sigma_c^2 \sin^2\bar{c} + \sigma_s^2 \cos^2\bar{c}$ and $\Sigma_{v_x v_y} = \Sigma_{v_y v_x} = (\sigma_s^2 - \bar{s}^2\sigma_c^2) \sin\bar{c} \cos\bar{c}$.

9.7.2 Performance criteria

As in the previous chapters, the performance analysis of various filtering techniques was done by obtaining RMSE, track-loss and computational time. The effect of initial uncertainty, measurement noise and sampling time on filter accuracy is also studied.

The root mean square error (RMSE) of range is defined as

$$RMSE_k = \sqrt{\frac{1}{M} \sum_{j=1}^{M} [(x_{j,k}^t - \hat{x}_{j,k}^t)^2 + (y_{j,k}^t - \hat{y}_{j,k}^t)^2]},$$

where k denotes the time steps and M, the total number of Monte Carlo runs.

A track is considered divergent and lost when the estimate moves away from the truth path without converging. So track-loss is considered in such conditions where the difference between true position and estimated position is more than a certain bound, which is taken as $1\ Km$ here. Hence track-loss is defined as $pos_{error} > 1\ Km$, where,

$$pos_{error} = \sqrt{(x_{k_{max}}^t - \hat{x}_{k_{max}}^t)^2 + (y_{k_{max}}^t - \hat{y}_{k_{max}}^t)^2},$$

and k_{max} is the final time step. Here, the reader may note that the passive tracking described here is not expected to run until the interception of the enemy ship. The target positions obtained from target motion analysis (TMA) are fed to the fire control system, which again tracks the target in active mode so that a possible interception is achieved.

9.7.3 Performance analysis of Gaussian sum filters

For simulation, sampling time was taken as $T = 1min$ and the total observation period lasted for $30min$. The accuracy level for SGHF and GS-SGHF is taken as 2 and the univariate points considered for GHF and GS-GHF are 3 in number. The displacement parameter used for splitting a multivariate Gaussian distribution is taken as 0.5, where the density is split into a total of 3 Gaussian components [44]. Figures 9.4 and 9.5 show the tracking performance of GS-SRF for both the cases, where the estimated target path is plotted with the truth target path for a single run. Here, Figure 9.4 is for case 1 and Figure 9.5 is for case 2. From these figures, it can be observed that GS-SRF tracks the target.

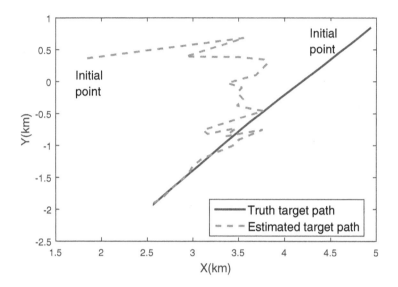

FIGURE 9.4: Truth and estimated trajectory (GS-SRF) for the first scenario [124].

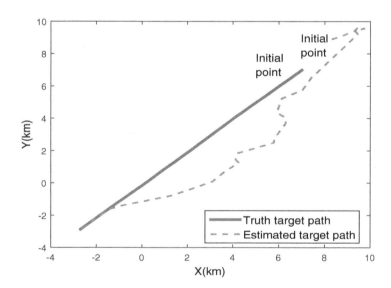

FIGURE 9.5: Truth and estimated trajectory (GS-SRF) for the second scenario [124].

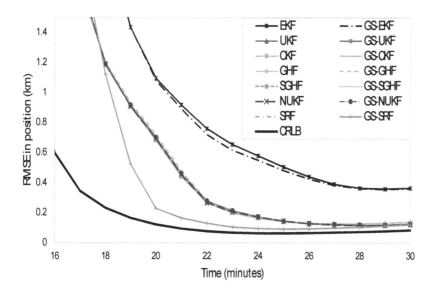

FIGURE 9.6: RMSE of range obtained from various filters at the first scenario.

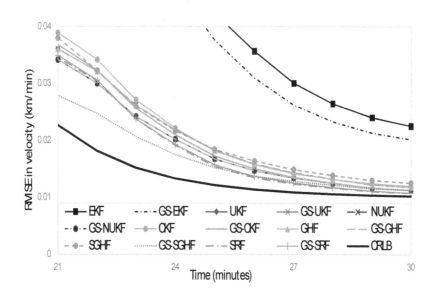

FIGURE 9.7: RMSE of velocity obtained from various filters at the first scenario [124].

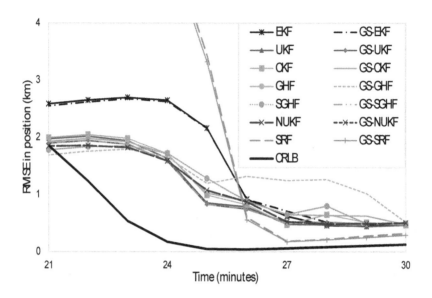

FIGURE 9.8: RMSE of range obtained from various filters at the second scenario [124].

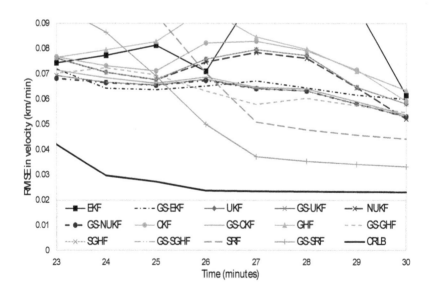

FIGURE 9.9: RMSE of resultant velocity obtained from various filters at the second scenario [124].

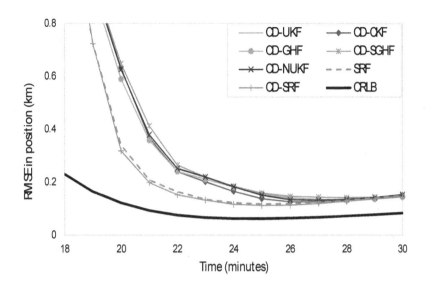

FIGURE 9.10: RMSE of range with CD filters for the scenario 1 [125].

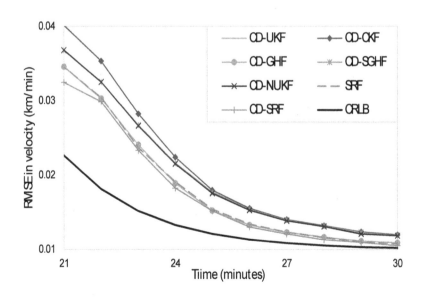

FIGURE 9.11: RMSE of velocity with CD filters for scenario 1 [125].

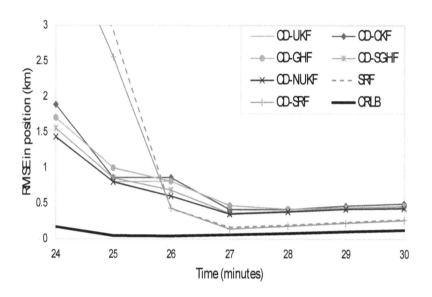

FIGURE 9.12: RMSE of range with CD filters for the second scenario [125].

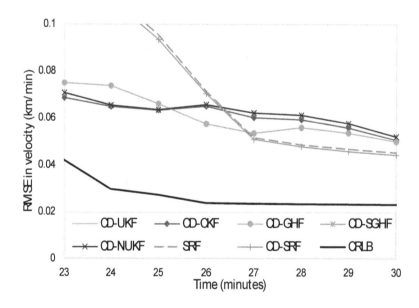

FIGURE 9.13: RMSE of resultant velocity with CD filters for the second scenario [125].

TABLE 9.2: % of track-loss for Gaussian sum filters w.r.t $\bar{\mu}$, σ_θ and T for case 1 [124].

		EKF	UKF	CKF	GHF	SGHF	GS-EKF	GS-UKF	GS-CKF	GS-GHF	GS-SGHF	NUKF	GS-NUKF	SRF	GS-SRF
$\bar{\mu}$	1	0.84	0.54	0.58	0.41	0.53	0.25	0.12	0.09	0.1	0.12	0.54	0.11	0	0
	2	1.43	0.71	0.67	0.55	0.71	0.25	0.13	0.11	0.13	0.12	0.59	0.13	0	0
	3	2.2	0.94	0.95	0.69	0.92	0.28	0.17	0.2	0.16	0.17	0.87	0.18	0	0
	4	2.87	1.22	1.26	0.96	1.22	0.34	0.21	0.28	0.19	0.2	1.09	0.19	0	0
	5	3.64	1.52	1.47	1.06	1.52	0.4	0.29	0.31	0.26	0.27	1.50	0.25	0	0
σ_θ	1	0.54	0.33	0.15	0.23	0.33	0.23	0.09	0.02	0.08	0.09	0.31	0.09	0	0
	1.5	0.84	0.54	0.58	0.41	0.53	0.25	0.12	0.09	0.1	0.12	0.54	0.11	0	0
	2	2.51	1.28	1.12	0.98	1.28	0.63	0.39	0.26	0.35	0.36	1.12	0.35	0	0
	2.5	3.04	1.59	1.11	1.22	1.59	0.78	0.43	0.31	0.45	0.43	1.35	0.40	0	0
	3	3.82	1.68	1.51	1.52	1.68	0.92	0.54	0.35	0.49	0.54	1.55	0.51	0	0
	3.5	4.42	1.89	1.73	1.76	1.89	1.22	0.57	0.39	0.53	0.55	1.78	0.55	0	0
	4	5.36	2.42	2.24	2.2	2.4	1.36	0.58	0.54	0.57	0.59	2.25	0.60	0.01	0
T	1	0.84	0.54	0.58	0.41	0.53	0.25	0.12	0.09	0.1	0.12	0.54	0.11	0	0
	1.5	7.57	1.4	1.5	2.2	1.3	7.07	0.8	0.7	0.9	0.6	1.4	0.7	0	0
	2	29.3	2.6	2.2	5.1	2.4	27.8	1.8	1.7	3.2	1.9	2.4	1.7	0	0
R.c.t		1	2.1	2.1	6.6	2.2	1.5	2.9	2.9	15.7	3.2	2.5	3.6	2.1	3.3

To compare the estimation accuracy of various filters, the RMSE in position and velocity for both the cases are plotted in Figures 9.6, 9.7, 9.8 and 9.9 where the diverged tracks are excluded (as per the definition for track-loss in this chapter). Figures 9.6 and 9.7 are for case 1 and it was observed that convergence starts only after observer manoeuvres. For case 1, the EKF and its Gaussian sum formulation performed with less accuracy whereas deterministic sample point filters and their Gaussian sum extensions performed with higher and comparable accuracy. On the other hand, SRF and GS-SRF performed more accurately than all the quadrature based filters and their Gaussian sum extensions. It was noted that GS-SRF converges much faster with a terminal error of $119m$. The corresponding error for SRF was $122m$, with all the sigma point filters giving an accuracy of around $140m$. However, the RMSE of GS-SRF is still a bit higher than the Cramer Rao lower bound (CRLB).

The RMSE in position and velocity after excluding the diverged tracks for case 2 are plotted in Figures 9.8 and 9.9. Since this is a highly nonlinear scenario, it takes more time for the filters to converge after observer manoeuvres. In this case also, GS-SRF performed with higher accuracy which is comparable with SRF, in the case of RMSE in position. But in the case of RMSE in velocity, GS-SRF performed with superior estimation accuracy as shown in Figure 9.9. Here, the terminal error given by GS-SRF is $288m$ while the corresponding error for SRF was $320m$. The terminal error for all the sigma point filters ranged between $500m$ to $700m$. Here also, the RMSE of GS-SRF is higher than CRLB. Another important thing that has to be noted in Figure 9.8 is the low RMSE values of all filters except SRF and GS-SRF in the initial simulation period. For a few time instants, RMSE values of these filters can even be found lower than CRLB too. This is due to the fact that in case 2, all filters other than SRF and GS-SRF incur very high track-loss (see Table 9.3). Since RMSE values are calculated after excluding track-loss instances, the erroneous tracks are omitted. Owing to this reason, these filters incur fewer RMSE values which are even lower than CRLB for a few time instants.

The tracking efficiency of GS-SRF is further explored in terms of track-loss as shown in Table 9.2 for case 1, where the effect of initial covariance, measurement noise covariance and sampling time on track-loss is studied. To study the effect of initial covariance, $\Sigma_{0|0} = \bar{\mu}\Sigma_{0|0}$ is assumed. It can be noted that the number of diverged tracks steadily increased with increase in $\bar{\mu}$ for all the deterministic sample point filters. For the same case, SRF and GS-SRF performed with superior tracking accuracy with not even a single diverged track. Table 9.2 also gives results of % track-loss when the measurement noise, σ_θ is varied. From these results, it can be observed that the tracking performance of all deterministic sample point filters suffered when the measurements are highly noise corrupted. However, GS-SRF performed with superior accuracy with zero number of track-loss, whereas for $\sigma_\theta = 4^o$, SRF incurred a single diverged track.

An increase in the sampling time resulted in a decrease in the accuracy for all the filters except SRF and GS-SRF, as shown in Table 9.2. All the quadrature filters incurred high track-loss whereas SRF and GS-SRF did not even lose a single track. From the results in Table 9.2, it can be observed that out of all the Gaussian filters, EKF gave the worst filtering performance while other deterministic sampling point filters performed with comparable accuracy. The main highlight of the results presented in Table 9.2 is the filtering accuracy of SRF and the proposed GS-SRF. It can be observed that GS-SRF performed with zero track-loss irrespective of an increase in initial uncertainty, measurement noise and sampling time. Relative computational time (R.c.t) for all the filters is listed in Table 9.2. From this, it can be noted that the computational load of GS-SRF is comparable to other Gaussian sum filters.

Table 9.3 shows the superior filtering accuracy of GS-SRF, when case 2 is considered. An increase in initial covariance resulted in an almost complete track-loss for all the quadrature filters whereas GS-SRF suffered from a maximum track-loss of only 5%, which is less than SRF track-loss. The effect of the increase in measurement noise also resulted in very high track-loss for quadrature filters and their Gaussian sum extensions. It can be noted that for small values of σ_θ, SRF and GS-SRF incurred almost the same number of diverged tracks. However, for higher values, the filtering performance of GS-SRF is more accurate such that the number of diverged tracks was lower when compared to SRF. Similar kinds of performance were observed for all the filters when sampling time was increased, as mentioned in Table 9.3. From all these results, it can be noted that except for SRF and GS-SRF, all other Gaussian and Gaussian sum filters failed with unacceptable track-loss percentage. Out of the SRF and the GS-SRF, the latter performed with higher track accuracy.

TABLE 9.3: % of track-loss for Gaussian sum filters w.r.t $\bar{\mu}$, σ_θ and T for case 2 [124].

		EKF	UKF	CKF	GHF	SGHF	GS-EKF	GS-UKF	GS-CKF	GS-GHF	GS-SGHF	NUKF	GS-NUKF	SRF	GS-SRF
$\bar{\mu}$	1	77.2	72	75.6	74.2	72.3	73.7	70.5	73.2	75.9	73.6	71.6	70.7	1	0.8
	2	85.6	83.8	82.7	82.3	83.2	84.2	81.6	80.9	80.2	81.7	83.4	81.2	1.4	1.1
	3	88.4	86.5	85.8	85.3	86	87.9	82.8	82.4	82.1	82.4	86.6	82.5	4	3.4
	4	90.1	89.2	89	88.7	89.8	89.2	86.7	86.2	86	86.5	89	86.4	5.7	4.3
	5	92.9	91.7	90.1	89.5	91.4	91.6	90	89.4	88.1	89.6	91.4	81.3	7.9	4.8
	1	56.5	37.1	36.4	35.5	37.3	54.2	33.1	33.9	37.3	33.1	36.3	33.2	0.3	0.1
	1.5	68.1	60.1	58.3	59.2	59.7	67.4	56.9	55.8	58.7	56.4	59.4	56.4	0.5	0.5
	2	77.2	72	75.6	74.2	72.3	73.7	70.5	73.2	75.9	73.6	71.6	70.7	1	0.8
σ_θ	2.5	82.9	79.8	81.1	78.2	80.2	77.3	75.8	76.1	75.1	75.8	79.8	75.9	2	1.3
	3	84.2	83.9	82.4	83.1	83.5	80.4	79.1	80.2	79.4	79.9	83.5	79.5	2.5	2.2
	3.5	87.5	85.1	84.7	83.9	84.7	83.8	81.7	82.1	81.3	81.8	84.6	81.9	4.9	2.8
	4	89.6	87.8	87.4	88.2	89.2	88.2	86.2	85.4	87.7	86.5	87.4	86.2	7.8	3.4
T	1	77.2	72	75.6	74.2	72.3	73.7	70.5	73.2	75.9	73.6	71.6	70.7	1	0.8
	1.5	82	79.3	76.2	80.4	79.1	80.2	79.8	76.6	81.6	75.1	79.2	79.1	3.5	1.2
	2	84.8	81.2	81.2	81.9	81.4	81.8	80.8	78.1	82.8	79.2	81.3	80.6	6.4	3.9

9.7.4 Performance analysis of continuous-discrete filters

All the filters were initialized according to the method described in the above Section 9.7.3. The performance of CD-SGHF, CD-NUKF and CD-SRF has been compared with other available CD filters. In this study, two observer manoeuvring scenarios are considered as given in Section 9.3.1. The performance has been compared in terms of RMSE, track-loss and computational time. Moreover, a parametric study has also been conducted to further explore the accuracy of CD filters.

The RMSE in position and velocity for all the CD filters for case 1 and case 2 are shown in Figures 9.10, 9.11, 9.12 and 9.13. Here also, diverged tracks are excluded according to the track-loss condition mentioned earlier. From Figure 9.10, where RMSE in position for case 1 is illustrated, it can be observed that the SRF and proposed CD-SRF converge much faster than other deterministic sample point filters. Among all the filters, CD-SRF performed with accuracy superior to all other CD filters, with a terminal error of $108m$. The corresponding error for SRF was $122m$. The terminal error incurred by the proposed CD-SGHF and CD-NUKF was $119m$ and $128m$. The RMSE in velocity for case 1 is shown in Figure 9.11, where a similar kind of performance with respect to RMSE in position is obtained.

Similarly, the RMSE in position for case 2 is plotted in Figure 9.12. As this is a highly nonlinear scenario, it takes more time for the filters to converge after observer manoeuvres. In this case also, CD-SRF performed with higher accuracy which is comparable with SRF. Here, the terminal error given by CD-SRF is comparable with that of GS-SRF at a value of $285m$. For all other filters also, terminal error was comparable to that of their Gaussian sum counterparts. From Figure 9.13, where RMSE in velocity is shown, it can be observed that SRF and CD-SRF performed with improved accuracy over all other filters, with CD-SRF giving a slightly improved result.

TABLE 9.4: % of track-loss for CD filters w.r.t $\bar{\mu}$, σ_θ and T for case 1 [124].

		UKF	CKF	GHF	SGHF	NUKF	SRF	CD-UKF	CD-CKF	CD-GHF	CD-SGHF	CD-NUKF	CD-SRF
$\bar{\mu}$	1	0.54	0.58	0.41	0.53	0.52	0	0.5	0.49	0.41	0.4	0.44	0
	2	0.71	0.67	0.55	0.71	0.68	0	0.62	0.63	0.53	0.58	0.59	0
	3	0.94	0.95	0.69	0.92	0.92	0	0.76	0.8	0.67	0.74	0.75	0
σ_θ	1.5	0.54	0.58	0.41	0.53	0.52	0	0.5	0.49	0.41	0.4	0.44	0
	2	1.28	1.12	0.98	1.28	1.18	0	1.19	0.97	0.91	1.17	1.03	0
	3	1.68	1.51	1.52	1.68	1.58	0	1.62	1.49	1.76	1.59	1.51	0
	4	2.42	2.24	2.2	2.4	2.37	0.1	1.93	1.92	2.1	1.9	1.87	0
T	1	0.54	0.58	0.41	0.53	0.52	0	0.5	0.49	0.41	0.40	0.44	0
	1.5	1.4	1.5	2.2	1.3	1.3	0	1.31	1.21	1.85	1.19	1.28	0
	2	2.6	2.2	5.1	2.4	2.5	0	2.2	2.02	5.4	2.1	2.14	0
R.c.t.		0.06	0.05	0.34	0.06	0.07	0.06	0.73	0.69	1	0.73	0.78	0.69

To further explore the tracking efficiency and robustness of CD-SRF, a parametric study was conducted where the effect of increasing initial uncertainty, measurement noise covariance and sampling time was considered. To study the effect of initial uncertainty, the initial covariance $\Sigma_{0|0}$ is varied according to the relation considered in the earlier section. Tracking accuracy of the proposed filters is explored by counting the number of diverged tracks. The percentage of track-loss incurred by all the filters for case 1 is listed in Table 9.4. It can be noted that the number of diverged tracks steadily increased with increase in $\bar{\mu}$ for all the quadrature filters. For the same case, SRF and CD-SRF performed with superior tracking accuracy with not even a single diverged track. Table 9.4 also gives results of % track-loss when the measurement noise, σ_θ is varied. From these results, it can be noted that the tracking performance of quadrature filters suffered when the measurements are highly noise corrupted. However, CD-SRF performed with superior accuracy with zero number of track-loss, whereas for $\sigma_\theta = 4^o$, SRF incurred a single diverged track. However, the performance of CD-SGHF and CD-NUKF was comparable to other quadrature based CD filters.

Increase in sampling time resulted in a decrease in accuracy for all the filters except SRF and CD-SRF, as shown in Table 9.4. All the deterministic sample point filters incurred higher track-loss whereas SRF and CD-SRF did not lose even a single track. From the results in Table 9.4, it can be observed that out of all the filters, deterministic sample point filters in the continuous-discrete domain performed with comparable accuracy. The main highlight of the results presented in Table 9.4 is the filtering accuracy of the CD-SRF. It can be observed that CD-SRF performed with zero track-loss irrespective of increase in initial uncertainty, measurement noise and sampling time. Relative computational time for all the filters is listed in Table 9.4. From this, it can be noted that the computational cost of CD-SRF, CD-NUKF and CD-SGHF is comparable to other CD quadrature filters.

TABLE 9.5: % of track-loss for CD filters w.r.t $\bar{\mu}$, σ_θ and T for case 2 [124].

		UKF	CKF	GHF	SGHF	NUKF	SRF	CD-UKF	CD-CKF	CD-GHF	CD-SGHF	CD-NUKF	CD-SRF
$\bar{\mu}$	1	72	75.6	74.2	72.3	71.7	1	53.4	53.5	51.6	53.1	52.9	0.6
	2	83.8	82.7	82.3	83.2	82.7	1.4	80.1	79.1	80	79.5	79.7	1.1
	3	86.5	85.8	85.3	86	86.1	4	84.5	83.5	85.1	83.3	83.6	2.86
σ_θ	1.5	60.1	58.3	59.2	59.7	59.4	0.5	44.2	42.4	37.9	40.17	43.4	0.22
	2	72	75.6	74.2	72.3	71.7	1	53.4	53.5	51.6	53.1	52.9	0.6
	3	83.9	82.4	83.1	83.5	82.7	2.5	72.8	72.7	74.56	70.1	71.5	2.1
	4	87.8	87.4	88.2	89.2	87.6	7.8	83.2	81.3	85.6	80.8	82.6	4.45
T	1	72	75.6	74.2	72.3	71.7	1	53.4	53.5	51.6	53.1	52.9	0.6
	1.5	79.3	76.2	80.4	79.1	77.4	3.5	65.8	66.9	69.2	65.3	66.2	0.9
	2	81.2	81.2	81.9	81.4	81	6.4	73.5	74.8	78.8	72	72.8	2.8

Table 9.5 shows the superior filtering accuracy of CD-SRF, where case 2 is considered. Increase in initial covariance resulted in an almost complete track-loss for all the CD deterministic sample point filters whereas CD-SRF suffered from a maximum track-loss of only 3%, which is less than SRF track-loss. The effect of an increase in measurement noise also resulted in very high track-loss for CD quadrature filters. It can be noted that for small values of σ_θ, SRF and CD-SRF incurred almost the same number of diverged tracks. However, for higher values, the filtering performance of CD-SRF is more accurate such that the number of diverged tracks was less when compared to SRF. A similar kind of performance was observed for all the filters when sampling time was increased, as mentioned in Table 9.5. From all these results, it can be noted that except for SRF and CD-SRF, all other CD filters and their discrete equivalents failed with an unacceptable track-loss percentage. Out of SRF and CD-SRF, the latter performed with higher tracking accuracy. From all these simulations, it can be inferred that the performance of GS-SRF and CD-SRF is comparable, and better than SRF and all other deterministic sample point filters considered.

9.8 Summary

In the above discussion, popular deterministic sample point filters, their continuous discrete and Gaussian sum versions are implemented for two typical BOT problems. Although SRF is not a deterministic sample point filter, the same has been implemented as it is specifically developed for a BOT problem and expected to perform better than all existing filters. Performance of the filters was studied by plotting the RMSE and comparing it with that of CRLB. Other parameters considered were the percentage of track-loss and computational time. The effect on tracking accuracy due to initial uncertainties, measurement noise covariance and sampling time is also studied. The findings of this work can be summarized as follows:

- Out of all the filters, GS-SRF and CD-SRF give the overall best result with higher accuracy, less track-loss and moderate computational time.

- Increase in initial uncertainty resulted in an increase in track-loss count for all the filters. GS-SRF and CD-SRF proved to be more robust and accurate than all other filters.

- Increasing the measurement noise resulted in lowering the filtering accuracy. Out of all the filters, GS-SRF and CD-SRF gave more accurate results.

- Varying the sampling time had a considerable effect on the number of diverged tracks. GS-SRF and CD-SRF gave the best result in this case with lower track-loss percentage.

To conclude, we request the reader to note that SRF is a heuristic designed *specifically* for the BOT problem, and it is no surprise that it outperforms other nonlinear filters. As expected, deterministic sampling point filters do outperform the EKF and their accuracy as well as percentage track loss is comparable with each other. Further, the GS-SRF and CD-SRF have potential to be an efficient tracking algorithm for the BOT problem with high tracking accuracy and moderate computational cost.

9.9 Tracking of a ballistic target

Solving a tracking problem is important to successfully intercept enemy missiles, satellite debris etc., and destroy them at a desired distance from the ground, by a guided missile [109, 147]. As a second case study, we consider tracking and interception of a ballistic target on reentry. Ballistic missiles, having a predetermined trajectory are usually targeted against civilian and industrial targets of strategic importance. The name ballistic target comes from the fact that the flight path is predetermined to a large extent according to the specific characteristics of the target [101].

The interception system for ballistic targets consists of a platform carrying an interceptor missile with an inbuilt seeker, propulsion unit, steering system, destruction capability system and processing device. All these components communicate with the processing device for a successful interceptor flight. For a possible interception, kinematic information of the target like position, velocity and acceleration are required, which is fed from a suitable tracking system. The entire flight duration of an interceptor missile can be divided into the following phases i) Initial guidance: where the interceptor is placed in such a way that the velocity vector is oriented in the direction of the target ii) Mid-course guidance: where the interceptor flight path is adjusted or oriented with the target direction so that it may facilitate an easy guidance in the final phase and iii) Terminal guidance: where the interceptor comes very near to the target so that it can go for a successful interception. In mid-course guidance, ground RADAR systems may send the measurements (obtained from the target) to the interceptor periodically, so that the guidance commands can be generated on-board. Terminal guidance requires very fast generation of guidance commands and hence, on-board tracking is necessary and this requires a seeker. From the seeker measurements, the tracking system generates the required estimates for the guidance algorithm. This guidance scheme is also called homing guidance. This is illustrated in Figure 9.14.

Successful interception of a ballistic target depends on the homing performance of the interceptor, that is, on the terminal guidance phase. The inbuilt seeker present in the interceptor acquires the noisy measurements which are functions of relative states of the target-interceptor model. These measurements are then processed by a state estimator for estimating the target kinematics. The guidance law receives these estimates and information about target position velocity etc., from GPS INS data fusion algorithm. The guidance block generates a guidance command based on the information on the target's and interceptor's position. Guidance commands are fed to the autopilot which creates deflections of the control surface and interceptor steers so that a successful interception can be achieved. A detailed block diagram is presented in Figure 9.15. Apart from other factors, accuracy of interception largely depends on the estimator's performance and guidance law [42]. Since both the process and measurement model considered for estimation are nonlinear, a highly accurate nonlinear filter is of great importance.

This section focuses on application of several advanced nonlinear filters for tracking ballistic targets, where the interceptor missile accelerations are defined using the proportional navigation guidance (PNG) law, since it has proved to be efficient for the terminal guidance phase [181]. For realizing the estimator block, deterministic support point, Gaussian sum, adaptive Gaussian sum and continuous-discrete filters have been employed. The performance of the closed-loop interceptor has been studied in terms of the RMSE of target accelerations and the final miss-distance. We shall see that overall, adaptive Gaussian sum filters performed with better accuracy, as compared to other filtering algorithms, incurring minimum miss-distance .

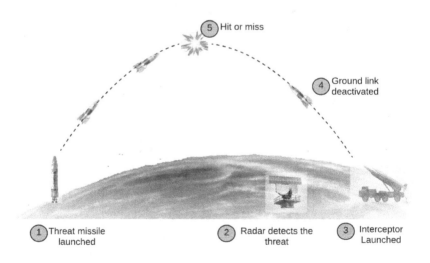

FIGURE 9.14: Schematic diagram of ballistic missile defence strategy.

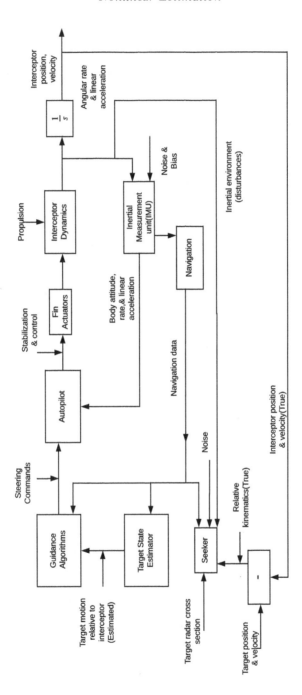

FIGURE 9.15: Block diagram of control, guidance and target tracking in-built in an interceptor.

9.10 Problem formulation

In this section, 3D behavior of a ballistic target and interceptor is discussed, where the measurements are obtained by an active seeker mounted on an interceptor missile. Here, a 6DOF SDC model of target-interceptor dynamics [42, 101] is adopted where the target accelerations are replaced by a single parameter, the ballistic coefficient (β) which is included as a state.

9.10.1 Process model

Let us assume x_t, y_t, z_t, v_{tx}, v_{ty}, v_{tz}, a_{tx}, a_{ty}, a_{tz} are the positions, velocities and accelerations of the target in x, y and z direction respectively. Similarly, x_m, y_m, z_m, v_{mx}, v_{my}, v_{mz}, a_{mx}, a_{my}, a_{mz} are the positions, velocities and accelerations of the interceptor missile. Now the relative positions and velocities in the inertial frame can be represented as

$$\Delta x = x_t - x_m, \quad \Delta y = y_t - y_m, \quad \Delta z = z_t - z_m,$$
$$\Delta v_x = v_{tx} - v_{mx}, \quad \Delta v_y = v_{ty} - v_{my}, \quad \Delta v_z = v_{tz} - v_{mz}.$$

The rate of change of relative positions and velocities can be written as [42]

$$\Delta \dot{x} = \Delta v_x, \ \Delta \dot{y} = \Delta v_y, \ \Delta \dot{z} = \Delta v_z, \ \Delta \dot{v}_x = a_{tx} - a_{mx},$$
$$\Delta \dot{v}_y = a_{ty} - a_{my}, \ \Delta \dot{v}_z = a_{tz} - a_{mz}, \ 1/\dot{\beta} = 0 \tag{9.12}$$

9.10.1.1 Process model in discrete domain

By discretizing equation (9.12) and incorporating various modeling errors, the process model becomes

$$\mathcal{X}_{k+1} = \mathcal{F}_k \mathcal{X}_k - GB_k + \eta_k, \tag{9.13}$$

where $\mathcal{X}_k = [\Delta x_k \ \Delta y_k \ \Delta z_k \ \Delta v_{x,k} \ \Delta v_{y,k} \ \Delta v_{z,k} \ 1/\beta_k]^T$, $\eta_k \sim \mathcal{N}(0, Q_k)$,

$$\mathcal{F}_k = \begin{bmatrix} I_{3\times3} & TI_{3\times3} & 0_{3\times1} \\ 0_{3\times3} & I_{3\times3} & [\mathbf{f}_1 \ \mathbf{f}_2 \ \mathbf{f}_3]^T \\ 0_{1\times3} & 0_{1\times3} & 1 \end{bmatrix},$$

$G = [T^2/2I_{3\times3} \ TI_{3\times3} \ 0_{1\times3}]^T$ and $B = [a_{mx_k} + g \ a_{my_k} \ a_{mz_k}]^T$, with $\mathbf{f}_1 = -0.5\bar{\rho}_k v_{t,k} v_{tx,k} T$, $\mathbf{f}_2 = -0.5\bar{\rho}_k v_{t,k} v_{ty,k} T$ and $\mathbf{f}_3 = -0.5\bar{\rho}_k v_{t,k} v_{tz,k} T$. Here, v_t is the resultant target velocity, T is the sampling instant and $\bar{\rho}$ is the air density which depends on the altitude of the target, expressed as

$$\bar{\rho} = 0.36392 \exp(-0.0001516584(x_t - 11000)) \ x_t > 11km$$
$$= 1.225(1 - 0.000022557695x_t)^{4.25587} \qquad x_t < 11km.$$

Here, g is the acceleration due to gravity which is considered to be a constant irrespective of the altitude.

9.10.1.2   Process model in continuous time domain

The stochastic differential equation representing the continuous process model can be described as

$$dX(t) = f(X(t), t)dt + \sqrt{Q}d\varsigma(t), \tag{9.14}$$

where $f(X)$ is as mentioned in Eq. (9.12). Using the Ito-Taylor expansion of order 1.5 for the time interval $(t, t + \delta)$, the continuous process equation is approximated as

$$X_k^{j+1} = f_d(X_k^j) + \sqrt{Q}\eta + \mathbb{L}\eta', \tag{9.15}$$

where $f_d(X_k^j) = \mathcal{F}_k^j X_k^j - GB_k$, X_k^j denotes $X(t)$ at time $t = t_k + j\delta$, $1 \leq j \leq m_{step}$ and $\delta = T/m_{step}$. The matrix

$$\mathcal{F}_k^j = \begin{bmatrix} \mathbf{I}_{3\times3} & \delta\mathbf{I}_{3\times3} & \mathbf{0}_{3\times1} \\ \mathbf{0}_{3\times3} & \mathbf{I}_{3\times3} & [\mathbf{f}_1 \ \mathbf{f}_2 \ \mathbf{f}_3]^T \\ \mathbf{0}_{1\times3} & \mathbf{0}_{1\times3} & 1 \end{bmatrix},$$

$G = [\delta^2/2\mathbf{I}_{3\times3} \ \delta\mathbf{I}_{3\times3} \ \mathbf{0}_{1\times3}]^T$ with $\mathbf{f}_1 = -0.5\bar{\rho}_k v_{t,k} v_{tx,k}\delta$, $\mathbf{f}_2 = -0.5\bar{\rho}_k v_{t,k} v_{ty,k}\delta$ and $\mathbf{f}_3 = -0.5\bar{\rho}_k v_{t,k} v_{tz,k}\delta$, and B is the same as mentioned earlier. Matrix $\mathbb{L}$ can be derived as

$$\mathbb{L} = \begin{bmatrix} \mathbf{0}_{3\times3} & diag([\sqrt{Q}_{4,4} \ \sqrt{Q}_{5,5} \ \sqrt{Q}_{6,6}]) & \mathbf{0}_{3\times1} \\ \mathbf{0}_{3\times3} & \mathbf{0}_{3\times3} & \mathbf{0}_{3\times1} \\ \mathbf{0}_{1\times3} & \mathbf{0}_{1\times3} & 0 \end{bmatrix},$$

where $\sqrt{Q}_{i,j}$ denotes the square root of the $(i, j)^{th}$ element of matrix Q. Hence, process update is done using Eq. (9.15) with respect to δ which is much smaller than the measurement sampling time T.

9.10.2   Seeker measurement model

A seeker mounted on the interceptor measures range rate, gimbal angles in yaw and pitch plane. Range rate is considered in the LOS frame, gimbal angles along the fin-dash frame and LOS rates along the inner gimbal frame.

In Figure 9.16, the LOS frame of reference is represented by (X_l, Y_l, Z_l) axes and (X_i, Y_i, Z_i) represents the inertial frame of reference. It can be noted that the LOS vector is along the direction of X_l. Here, λ_a is the azimuth angle and λ_e is the elevation. They are expressed as [42], [123]

$$\lambda_{a,k} = \tan^{-1}\left(\frac{\Delta y_k}{\Delta z_k}\right) \tag{9.16}$$

and

$$\lambda_{e,k} = \tan^{-1}\left(\frac{\Delta x_k}{\sqrt{\Delta y_k^2 + \Delta z_k^2}}\right). \tag{9.17}$$

The rate of change of these angles are

$$\dot\lambda_{a,k} = \frac{\Delta z_k \Delta v_{y,k} - \Delta y_k \Delta v_{z,k}}{\Delta y_k^2 + \Delta z_k^2} \tag{9.18}$$

and

$$\dot\lambda_{e,k} = \frac{\Delta v_{x,k}(\Delta y_k^2 + \Delta z_k^2) - \Delta x_k(\Delta y_k \Delta v_{y,k} + \Delta z_k \Delta v_{z,k})}{(\Delta x_k^2 + \Delta y_k^2 + \Delta z_k^2)\sqrt{\Delta y_k^2 + \Delta z_k^2}}. \tag{9.19}$$

Now range rate, which is a measurement, can be expressed as

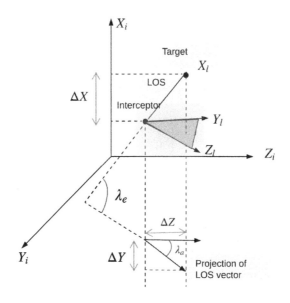

FIGURE 9.16: Representation of LOS frame with respect to inertial frame [126].

$$\dot r_k = \frac{\Delta x_k \Delta v_{x,k} + \Delta y_k \Delta v_{y,k} + \Delta z_k \Delta v_{z,k}}{\sqrt{\Delta x_k^2 + \Delta y_k^2 + \Delta z_k^2}}. \tag{9.20}$$

The measurements, gimbal angles along the yaw and pitch plane represented in the fin-dash frame can be expressed as $\phi_{y,k} = \tan^{-1}\left(\frac{-n_{fd,k}}{l_{fd,k}}\right)$ and $\phi_{z,k} = \tan^{-1}\left(\frac{m_{fd,k}}{\sqrt{l_{fd,k}^2 + n_{fd,k}^2}}\right)$, where $(l_{fd,k}, m_{fd,k}, n_{fd,k})$ are unit vectors along the fin-dash frame. Range vector in LOS frame can be transformed to the fin-dash frame with the following transformation

$$\begin{bmatrix} l_{fd,k} \\ m_{fd,k} \\ n_{fd,k} \end{bmatrix} = D_f^{fd} D_b^f D_i^b D_l^i \begin{bmatrix} 1 \\ 0 \\ 0 \end{bmatrix}.$$

D_l^i, D_i^b, D_b^f and D_f^{fd} are the direction cosine matrices (DCM) used for various frame transformations. These can be expressed as

$$D_l^i = \begin{bmatrix} \sin \lambda_e & 0 & -\cos \lambda_e \\ \cos \lambda_e \sin \lambda_a & \cos \lambda_a & \sin \lambda_e \sin \lambda_a \\ \cos \lambda_e \cos \lambda_a & -\sin \lambda_a & \sin \lambda_e \cos \lambda_a \end{bmatrix},$$

$$D_b^f = \begin{bmatrix} 1 & 0 & 0 \\ 0 & \dfrac{1}{\sqrt{2}} & \dfrac{1}{\sqrt{2}} \\ 0 & -\dfrac{1}{\sqrt{2}} & \dfrac{1}{\sqrt{2}} \end{bmatrix}$$

and

$$D_f^{fd} = \begin{bmatrix} 1 & 0 & 0 \\ 0 & -1 & 0 \\ 0 & 0 & -1 \end{bmatrix},$$

where D_l^i, D_b^f and D_f^{fd} represent frame transformation from LOS to inertial, body to fin and fin to fin-dash, respectively. The frame transformation matrix from inertial to body frame, D_i^b, can be obtained by considering the body frame quaternion with respect to inertial. If we define the quaternion as $(\hat{q}, q_0) = q_1 + q_2 i + q_3 j + q_4 k$, then

$$D_i^b = \begin{bmatrix} q_1^2 + q_2^2 - q_3^2 - q_4^2 & 2(q_2 q_3 + q_4 q_1) & 2(q_2 q_4 - q_3 q_1) \\ 2(q_2 q_3 - q_4 q_1) & q_1^2 - q_2^2 + q_3^2 - q_4^2 & 2(q_3 q_4 + q_2 q_1) \\ 2(q_2 q_4 + q_3 q_1) & 2(q_3 q_4 - q_2 q_1) & q_1^2 - q_2^2 - q_3^2 + q_4^2 \end{bmatrix}.$$

Now the measurements, LOS rates along roll, yaw and pitch plane represented in the inner gimbal frame are

$$\begin{bmatrix} \omega_{x,k} \\ \omega_{y,k} \\ \omega_{z,k} \end{bmatrix} = D_{fd}^g D_f^{fd} D_b^f D_i^b D_l^i \begin{bmatrix} -\dot{\lambda}_{a,k} \sin \lambda_{e,k} \\ \dot{\lambda}_{e,k} \\ \dot{\lambda}_{a,k} \cos \lambda_{e,k} \end{bmatrix},$$

where

$$D_{fd}^g = \begin{bmatrix} \cos \phi_y \cos \phi_y & \sin \phi_z & -\cos \phi_y \sin \phi_y \\ -\cos \phi_y \sin \phi_z & \cos \phi_z & \sin \phi_y \sin \phi_z \\ \sin \phi_y & 0 & \cos \phi_y \end{bmatrix}$$

represents the direction cosine matrix (DCM) for frame transformation from fin-dash frame to inner gimbal frame and $\lambda_{a,k}$, $\dot{\lambda}_{a,k}$, $\lambda_{e,k}$ and $\dot{\lambda}_{e,k}$ are obtained from Eqs. (9.16)-(9.19). Out of these three body rates, $\omega_{y,k}$ and $\omega_{z,k}$ are available as measurements. Now, the set of seeker measurements corrupted with sensor noises can be represented as

$$\dot{r}_k = \frac{\Delta x_k \Delta V_{x,k} + \Delta y_k \Delta V_{y,k} + \Delta z_k \Delta V_{z,k}}{\sqrt{\Delta x_k^2 + \Delta y_k^2 + \Delta z_k^2}} + v_{1,k} ,$$

$$\phi_{y,k} = \tan^{-1}\left(\frac{-n_{fd,k}}{l_{fd,k}}\right) + v_{2,k} ,$$

$$\phi_{z,k} = \tan^{-1}\left(\frac{m_{fd,k}}{\sqrt{l_{fd,k}^2 + n_{fd,k}^2}}\right) + v_{3,k} , \qquad (9.21)$$

$$\begin{bmatrix} \omega_{y,k} \\ \omega_{z,k} \end{bmatrix} = \begin{bmatrix} 0 & 1 & 0 \\ 0 & 0 & 1 \end{bmatrix} \begin{bmatrix} \omega_{x,k} \\ \omega_{y,k} \\ \omega_{z,k} \end{bmatrix} + \begin{bmatrix} v_{4,k} \\ v_{5,k} \end{bmatrix} ,$$

where $v_k = [v_{1,k}\ v_{2,k}\ v_{3,k}\ v_{4,k}\ v_{5,k}]^T$, and $v_k \sim \mathcal{N}(0, R)$.

9.10.3 Target acceleration model

Accurate modeling of target acceleration can be done by considering the drag force and gravity [101]. As we consider a free falling target, its acceleration can be represented as [123]:

$$\begin{bmatrix} a_{tx} \\ a_{ty} \\ a_{tz} \end{bmatrix} = \begin{bmatrix} a_{tx}^{drag} \\ a_{ty}^{drag} \\ a_{tz}^{drag} \end{bmatrix} + \begin{bmatrix} -g \\ 0 \\ 0 \end{bmatrix} ,$$

or $\bar{a} = \dfrac{\bar{D}}{m_t} + \bar{g}$, where $a_{tx}^{drag}, a_{ty}^{drag}$ and a_{tz}^{drag} are target acceleration components due to drag force in the x, y and z axis, respectively. The magnitude of drag force can be defined as $D = \mathsf{q} S_t C_d$, where $\mathsf{q} = 0.5\bar{\rho}v_t^2$ is the dynamic pressure, S_t is the cross sectional area of the ballistic target and C_d is the zero lift drag which is the function of speed of the ballistic target and its aerodynamic shape. Now ballistic coefficient β can be defined as $\beta = \dfrac{m_t}{S_r C_d}$, where m_t is the mass of the target. Using this relation, drag force can be written as $D = \dfrac{0.5\bar{\rho}v_t^2}{\beta}$.

Target acceleration components in the inertial frame can be defined as [42]

$$a_{tx,k} = \frac{-0.5\bar{\rho}_k v_{t,k} v_{tx,k}}{\beta_k} - g,$$

$$a_{ty,k} = \frac{-0.5\bar{\rho}_k v_{t,k} v_{ty,k}}{\beta_k}, \qquad (9.22)$$

$$a_{tz,k} = \frac{-0.5\bar{\rho}_k v_{t,k} v_{tz,k}}{\beta_k},$$

where the resultant target velocity, $v_{t,k} = \sqrt{v_{tx,k}^2 + v_{ty,k}^2 + v_{tz,k}^2}$. Here, $v_{tx,k} = \Delta v_{x,k} + v_{mx,k}$, $v_{ty,k} = \Delta v_{y,k} + v_{my,k}$ and $v_{tz,k} = \Delta v_{z,k} + v_{mz,k}$. $\Delta v_{x,k}$, $\Delta v_{y,k}$ and $\Delta v_{z,k}$ are obtained from the estimates and $v_{mx,k}$, $v_{my,k}$ and $v_{mz,k}$ are known (our own vehicle velocities).

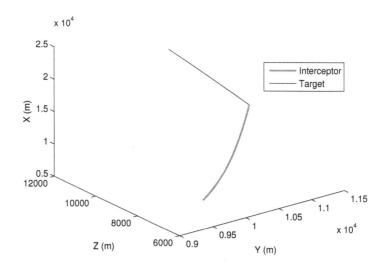

FIGURE 9.17: Target and interceptor trajectory [126].

TABLE 9.6: Initial truth values for relative states, $\mathcal{X}_{0|0}$ and initial uncertainty, $\Sigma_{0|0}$ [126].

States	$\mathcal{X}_{0|0}$	$\Sigma_{0|0}$
Δx	$14,296m$	$100m^2$
Δy	$846m$	$100m^2$
Δz	$4,280m$	$100m^2$
Δv_x	$1,946m/s$	$10(m/s)^2$
Δv_y	$262m/s$	$10(m/s)^2$
Δv_z	$662m/s$	$10(m/s)^2$
$1/\beta$	$1/3000$	$8/10^5$

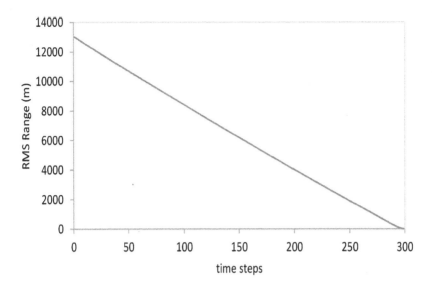

FIGURE 9.18: Root mean square of range.

9.11 Proportional navigation guidance (PNG) law

Here, the estimated states are used by guidance block, which is implemented with the proportional navigation guidance (PNG) law. The PNG law defines acceleration commands perpendicular to the target-interceptor line of sight (LOS) and are proportional to the rate of change of relative range between the target-interceptor dynamics, i.e., the closing velocity and the LOS rate. It can be mathematically expressed as $Ac = C'V_c\dot{\lambda}$, where Ac is the acceleration command, C' is the effective navigation ratio which is a constant, V_c is the closing velocity and $\dot{\lambda}$ is the LOS rate.

The extension of this guidance law to three dimensions is explained using the concept of zero effort miss ($\bar{\Lambda}$) [181]. $\bar{\Lambda}$ can be defined as the relative distance by which the target and missile would miss without an interception if both continue to follow their course without any change in velocity. The relative estimated states available from the nonlinear filter i.e., $\Delta x_k, \Delta y_k, \Delta z_k, \Delta v_{x,k}, \Delta v_{y,k}$ and $\Delta v_{z,k}$ are directly made use of for defining the interceptor missile acceleration commands in 3D with the help of a parameter t_{go} (time left for attaining a possible interception). For the PNG law, $\bar{\Lambda}$ vector component perpendicular to the LOS has to be computed and for this, we first compute $\bar{\Lambda}$ vector component parallel to LOS. As a general representation, $\bar{\Lambda}$ vector can be expressed as $\bar{\Lambda} = \bar{\Lambda}_x\,\hat{i} + \bar{\Lambda}_y\,\hat{j} + \bar{\Lambda}_z\,\hat{k}$, where $\bar{\Lambda}_x = \Delta x + \Delta v_x t_{go}$,

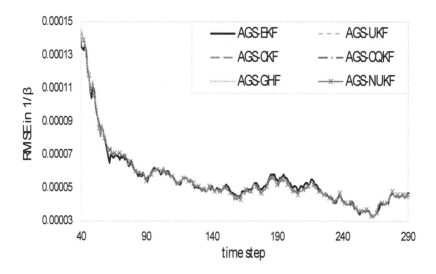

FIGURE 9.19: RMSE of $1/\beta$ with AGS filters.

$\bar{\Lambda}_y = \Delta y + \Delta v_y t_{go}$ and $\bar{\Lambda}_z = \Delta z + \Delta v_z t_{go}$. A unit vector in the direction of LOS will be $\hat{L} = (\Delta x \hat{i} + \Delta y \hat{j} + \Delta z \hat{k})/(\sqrt{\Delta x^2 + \Delta y^2 + \Delta z^2}) = (\Delta x \hat{i} + \Delta y \hat{j} + \Delta z \hat{k})/r$. The $\bar{\Lambda}$ vector parallel to the LOS can be found out using the dot product of $\bar{\Lambda}$ with unit LOS and is in the direction of $\hat{L}$. So the $\bar{\Lambda}$ vector parallel to LOS, $\bar{\Lambda}_p$ is expressed as $\bar{\Lambda}_p = \mathbf{X}\hat{L}$, where $\mathbf{X} = (\bar{\Lambda}_x \Delta x + \bar{\Lambda}_y \Delta y + \bar{\Lambda}_z \Delta z)/r$.

Now the vector difference between $\bar{\Lambda}$ and $\bar{\Lambda}_p$ will give the $\bar{\Lambda}$ component perpendicular to the LOS, i.e., $\bar{\Lambda}_P = \bar{\Lambda}_{Px}\,\hat{i} + \bar{\Lambda}_{Py}\,\hat{j} + \bar{\Lambda}_{Pz}\,\hat{k}$, where $\bar{\Lambda}_P = \bar{\Lambda} - \bar{\Lambda}_p$. The components of $\bar{\Lambda}_P$ are computed as $\bar{\Lambda}_{Px} = \bar{\Lambda}_x - (\mathbf{X}\Delta x)/r$, $\bar{\Lambda}_{Py} = \bar{\Lambda}_y - (\mathbf{X}\Delta y)/r$, $\bar{\Lambda}_{Pz} = \bar{\Lambda}_z - (\mathbf{X}\Delta z)/r$. Since for the PNG law, acceleration commands are proportional to the perpendicular component of $\bar{\Lambda}$, i.e., $\bar{\Lambda}_P$ and inversely proportional to the square of time to go, t_{go}, the modified form in 3D can be given as

$$a_{mx_k} = \frac{C'\bar{\Lambda}_{Px,k}}{t_{go,k}^2}, \quad a_{my_k} = \frac{C'\bar{\Lambda}_{Py,k}}{t_{go,k}^2} \quad \text{and} \quad a_{mz_k} = \frac{C'\bar{\Lambda}_{Pz,k}}{t_{go,k}^2}. \quad (9.23)$$

9.12 Simulation results

The target-interceptor scenario described above is used to study the performance of the interceptor under different filtering schemes combined with PNG law. The performance comparison is done in terms of averaged miss-distance, RMSE of states, and execution time computed over several Monte Carlo runs.

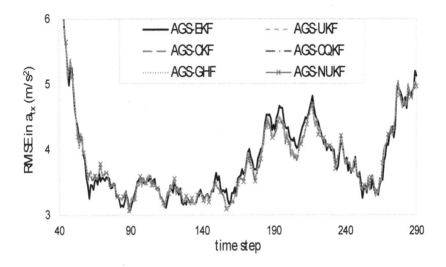

FIGURE 9.20: RMSE of a_{tx} in (m/s^2) with AGS filters.

In the absence of a real physical system, a truth model is generated virtually in the computer. The initial truth value for the interceptor is taken as $\mathcal{X}_{0|0} = [8690m \ 9460m \ 6340m \ 800m/s \ 363m/s \ 418m/s \ -7.84 \ m/s^2 \ - 60.31m/s^2 \ -23.27m/s^2]^T$. These are the values which are initially fed to the PNG law and after getting estimates from the filter, the interceptor position for the next time instant is defined. The initial estimate for the filter is assumed as $\mathcal{N}(\mathcal{X}_{0|0}, \Sigma_{0|0})$, where $\mathcal{X}_{0|0}$ and $\Sigma_{0|0}$ are given in Table 9.6. From these values, target states can be found out according to the relation $\mathcal{X}_{t,k} = \mathcal{X}_k + \mathcal{X}_{m,k}$. The synthetic measurement data is generated with the help of process and measurement equations. The covariance matrices for process and measurement noises are assumed as $Q = diag((0.02m)^2 \ (0.02m)^2 \ (0.02m)^2 \ (0.02m/s)^2 \ (0.02m/s)^2 \ (0.02m/s)^2 \ 4 \times 10^{-11})$ and $R = diag((15m)^2 \ (0.3°)^2 \ (0.3°)^2 \ (5°/s)^2 \ (5°/s)^2)$, respectively.

The initial pdf is split into a total of 3 components where each component is realized by the above mentioned filters. The weights of individual Gaussian components of AGS filters have been computed by solving the optimization problem mentioned. Seeker measurements are obtained at a rate of $T = 0.025s$ and the simulation period lasted for almost $7.25s$ (the simulation ends when the interceptor crosses the target). The crossing can be inferred from the change of direction of closing velocity, $V_{c,k}$.

Figure 9.17 shows the trajectory of the target and interceptor in three dimensions. The RMS of range (distance between target and interceptor) is plotted in Figure 9.18. This indicates the accurate closed-loop performance of the estimator and guidance law. However, the end error, which we call as

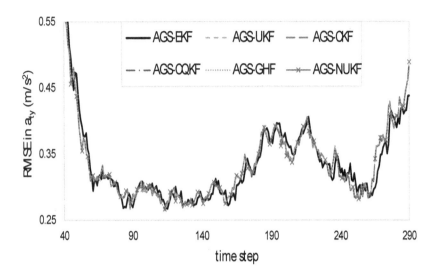

FIGURE 9.21: RMSE of a_{ty} in (m/s^2) with AGS filters.

miss-distance, needs to be evaluated closely for all the nonlinear filters clubbed
with the PNG law.

9.12.1   Performance of adaptive Gaussian sum filters

Here, adaptive Gaussian sum and Gaussian sum filters with NUKF, CQKF
and GHF proposals have been implemented for obtaining the solution. The
accuracy levels of CQKF and GHF are taken as $n' = N = 3$. The total number
of sample point requirements has been listed in Table 9.7.

The RMSE of $1/\beta$ is calculated out of 50 MC runs and is plotted in
Figure 9.19. From the figure, it can be seen that $1/\beta$ converges to its true
value before $1/5^{th}$ of the total homing guidance time. The RMSEs of target accelerations in all the three dimensions are given in Figures 9.20, 9.21
and 9.22, respectively. These figures denote the convergence and tracking of
the proposed filters. However, from the figures we cannot claim superiority
of one filter over the other. Miss-distances of all the Gaussian and Gaussian
sum filters combined with the PNG law averaged over 50 MC runs, have been
tabulated in Table 9.7. From the table, it can be concluded that (i) all deterministic sample point filters provide a miss-distance of almost $20m$, which is
better than the EKF. (ii) The GSF with the deterministic sample point filter
proposal provides considerably less miss-distance, around $16m$ when compared
to ordinary Gaussian filters. This is again better than the miss-distance given
by GS-EKF, which is $22.26m$. (iii) The miss-distance for AGSF is further less
when compared to ordinary GSF and Gaussian filters. Here, it should be noted

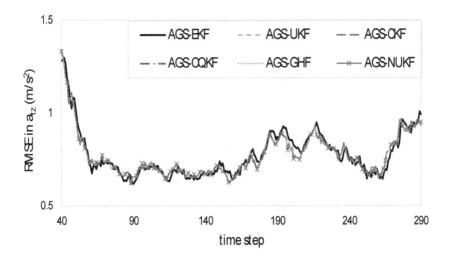

FIGURE 9.22: RMSE of a_{tz} in (m/s^2) with AGS filters.

that the proposed AGSF gave more accurate miss-distance values $(13.41m)$ when compared to the already existing AGS-EKF.

In Table 9.7, the relative computational times of different filters are listed. It can be seen that the slowest filter is AGSF with the GHF proposal and the fastest one is EKF. The execution time of AGSF, other than with the GHF proposal is comparable. Since AGSF with NUKF, CKF and CQKF performed with superior accuracy and moderate execution time, it is recommend to use any of these filters for tracking an incoming ballistic target.

9.12.2 Performance of continuous-discrete filters

The performance comparison is done in terms of averaged miss-distance, RMSE of states and execution time, computed over 100 Monte Carlo runs with $m_{step} = 5$. Since the 3D target-interceptor scenarios are similar to that of adaptive Gaussian sum filters shown above, they are avoided in this section. The RMSE of $1/\beta$ is plotted in Figure 9.23. It is important to accurately estimate this parameter since target accelerations are defined in terms of $1/\beta$, along with other relative states. From the figure, it can be observed that all CD filters performed with almost the same RMSE, which is much lower than that of discrete-discrete filters. The RMSE in a_{tx} is shown in Figure 9.24, from which it can be noted that all CD filters performed with comparable accuracy which is higher than their counterparts in the discrete domain. Similar performance can be observed for the RMSEs in a_{ty} and a_{tz} as shown in Figures 9.25 and 9.26.

Enhancement in filtering accuracy with CD filters is further illustrated by calculating the miss-distance, as given in Table. 9.8. From this table, it can

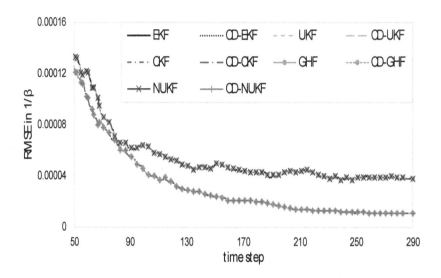

FIGURE 9.23: RMSE of $1/\beta$ with CD filters.

be observed that the CD quadrature filters performed with higher accuracy incurring a miss-distance of only $14m$, whereas CD-EKF incurred $17m$. For the ordinary discrete filters, it was found to be more which is almost $20m$ for quadrature filters and $22m$ for EKF. Relative computational time for all the filters is also listed in Table 9.8. It can be noted that CD-GHF takes the maximum computational time. The computational cost for all other CD filters was moderate and comparable. Hence it can be concluded that CD-NUKF or CD-CKF combined with the PNG law, which performed with comparable accuracy and moderate computational time, can be considered as a better solution for tracking and intercepting a ballistic target on reentry.

From all these results, it can be inferred that CD filters performed with more accuracy as they have low RMSE and comparable miss-distance with respect to Gaussian sum filters.

9.13   Conclusions

In the above section, ballistic target tracking and its interception on reentry are solved using various filters. For realizing the guidance algorithm, the PNG law is implemented in closed-loop with the nonlinear filter. Performance of the closed-loop interceptor is evaluated by calculating the RMSE in target-interceptor relative states, final miss-distance and computational time. From the RMSE plots, it can be observed that continuous-discrete filters performed

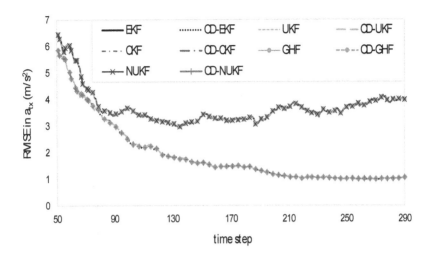

FIGURE 9.24: RMSE of a_{tx} in (m/s^2) with CD filters.

with somewhat less error than Gaussian and adaptive Gaussian sum filters, incurring a final miss-distance of around $14m$. For adaptive Gaussian sum filters, it was found that the RMSE incurred is similar to that of Gaussian filters while the final miss-distance was around $13m$. It has to be noted that among adaptive Gaussian sum filters, final miss-distance incurred by deterministic sample point filters is comparable and smaller than the EKF. All the Gaussian sum filters performed with a final miss-distance of around $16m$. Since continuous-discrete filters performed with moderate computational cost, lower RMSE and miss-distance, they can be considered for successfully intercepting a ballistic target on reentry.

TABLE 9.7: Miss-distance, number of points and relative computational time for AGS filters.

Filters	Miss-distance	No. points	Relative comp. time
AGS-NUKF	13.62m		0.91
GS-NUKF	14.75m		0.008
NUKF	21.33m	29	0.006
AGS-GHF	13.41m		1
GS-GHF	16.43m		0.73
GHF	20.61m	2187	0.33
AGS-CQKF	13.52m		0.10
GS-CQKF	16.13m		0.017
CQKF	20.36m	42	0.007
AGS-CKF	13.92m		0.080
GS-CKF	16.35m		0.006
CKF	20.52m	14	0.003
AGS-UKF	13.85m		0.082
GS-UKF	16.06m		0.006
UKF	20.39m	15	0.002
AGS-EKF	18.34m		0.077
GS-EKF	21.54m		0.004
EKF	22.26m		0.001

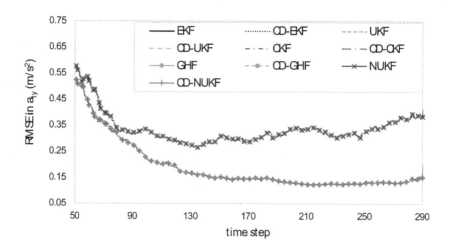

FIGURE 9.25: RMSE of a_{ty} in (m/s^2) with CD filters.

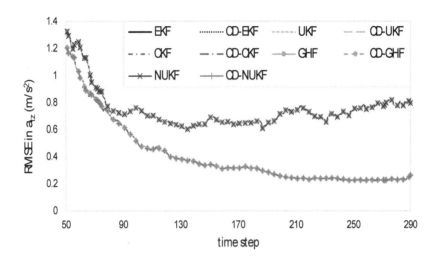

FIGURE 9.26: RMSE of a_{tz} in (m/s^2) with CD filters.

TABLE 9.8: Miss-distance and relative computational time for CD filters.

Filter	Miss-distance	Relative comp. time
EKF	$22.2m$	0.0014
CD-EKF	$17.5m$	0.0046
UKF	20.3	0.0029
CD-UKF	$14.8m$	0.0097
CKF	$20.5m$	0.0028
CD-CKF	$14.6m$	0.0091
GHF	$20.6m$	0.297
CD-GHF	$14.8m$	1
NUKF	$21.33m$	0.0033
CD-NUKF	$14.2m$	0.0012

Bibliography

[1] Seham Allahyani and Paresh Date. A minimum variance filter for continuous discrete systems with additive-multiplicative noise. In *Signal Processing Conference (EUSIPCO), 2016 24th European*, pages 2330–2334. IEEE, 2016.

[2] Brian D.O. Anderson and John B. Moore. *Optimal Filtering. Englewood Cliffs*, 21:22–95, 1979.

[3] Panos J. Antsaklis and Anthony N. Michel. *A Linear Systems Primer*, volume 1. Birkhäuser Boston, 2007.

[4] Ienkaran Arasaratnam and Simon Haykin. Square-root quadrature Kalman filtering. *IEEE Transactions on Signal Processing*, 56(6):2589–2593, 2008.

[5] Ienkaran Arasaratnam and Simon Haykin. Cubature Kalman filters. *IEEE Transactions on Automatic Control*, 54(6):1254–1269, 2009.

[6] Ienkaran Arasaratnam, Simon Haykin, and Robert J. Elliott. Discrete-time nonlinear filtering algorithms using Gauss–Hermite quadrature. *Proceedings of the IEEE*, 95(5):953–977, 2007.

[7] Ienkaran Arasaratnam, Simon Haykin, and Thomas R. Hurd. Cubature Kalman filtering for continuous-discrete systems: theory and simulations. *IEEE Transactions on Signal Processing*, 58(10):4977–4993, 2010.

[8] M. Sanjeev Arulampalam, Simon Maskell, Neil Gordon, and Tim Clapp. A tutorial on particle filters for online nonlinear/non-Gaussian Bayesian tracking. *IEEE Transactions on Signal Processing*, 50(2):174–188, 2002.

[9] M. Sanjeev Arulampalam, Branko Ristic, N. Gordon, and T. Mansell. Bearings-only tracking of manoeuvring targets using particle filters. *EURASIP Journal on Applied Signal Processing*, 2004:2351–2365, 2004.

[10] Michael Athans and Edison Tse. A direct derivation of the optimal linear filter using the maximum principle. *IEEE Transactions on Automatic Control*, 12(6):690–698, 1967.

[11] F. Auger, J. Guerrero, M. Hilairet, S. Katsura, E. Monmasson, and T. Orlowska-Kowalska. Introduction to the special section on industrial applications and implementation issues of the Kalman filter. *IEEE Transactions on Industrial Electronics*, 59(11):4165–4168, 2012.

[12] François Auger, Mickael Hilairet, Josep M. Guerrero, Eric Monmasson, Teresa Orlowska-Kowalska, and Seiichiro Katsura. Industrial applications of the Kalman filter: A review. *IEEE Transactions on Industrial Electronics*, 60(12):5458–5471, 2013.

[13] Yaakov Bar-Shalom, X. Rong Li, and Thiagalingam Kirubarajan. *Estimation with Applications to Tracking and Navigation: Theory Algorithms and Software*. John Wiley & Sons, 2004.

[14] V.E. Beneš. Exact finite-dimensional filters for certain diffusions with nonlinear drift. *Stochastics: An International Journal of Probability and Stochastic Processes*, 5(1-2):65–92, 1981.

[15] Shovan Bhaumik. Square-root cubature-quadrature Kalman filter. *Asian Journal of Control*, 16(2):617–622, 2014.

[16] Swati and Shovan Bhaumik. Nonlinear estimation using cubature quadrature points. In *Energy, Automation, and Signal (ICEAS), 2011 International Conference on*, pages 1–6. IEEE, 2011.

[17] Shovan Bhaumik and Swati. Cubature quadrature Kalman filter. *IET Signal Processing*, 7(7):533–541, 2013.

[18] Prova Biswas, Ashoke Sutradhar, and Pallab Datta. Estimation of parameters for plasma glucose regulation in type-2 diabetics in presence of meal. *IET Systems Biology*, 12(1):18–25, 2017.

[19] Andrew Blake and Michael Isard. The condensation algorithm-conditional density propagation and applications to visual tracking. *Conference on Advances in Neural Information Processing Systems*, December 2–5, pages 361–367, 1996.

[20] Robert Grover Brown and Patrick Y.C. Hwang. *Introduction to Random Signals and Applied Kalman Filtering: With MATLAB Exercises and Solutions, New York: Wiley, c1997.*, 1997.

[21] Hans-Joachim Bungartz and Michael Griebel. Sparse grids. *Acta Numerica*, 13:147–269, 2004.

[22] Daniela Calvetti, G. Golub, W. Gragg, and Lothar Reichel. Computation of Gauss-Kronrod quadrature rules. *Mathematics of Computation of the American Mathematical Society*, 69(231):1035–1052, 2000.

[23] JinDe Cao, R. Rakkiyappan, K. Maheswari, and A. Chandrasekar. Exponential H_∞ filtering analysis for discrete-time switched neural networks with random delays using sojourn probabilities. *Science China Technological Sciences*, 59(3):387–402, 2016.

[24] James Carpenter, Peter Clifford, and Paul Fearnhead. Improved particle filter for nonlinear problems. *IEE Proceedings-Radar, Sonar and Navigation*, 146(1):2–7, 1999.

[25] Goutam Chalasani and Shovan Bhaumik. Bearing only tracking using Gauss-Hermite filter. In *Industrial Electronics and Applications (ICIEA), 2012 7th IEEE Conference on*, pages 1549–1554. IEEE, 2012.

[26] K.P. Bharani Chandra, Da-Wei Gu, and Ian Postlethwaite. Square root cubature information filter. *IEEE Sensors Journal*, 13(2):750–758, 2013.

[27] Lubin Chang, Baiqing Hu, An Li, and Fangjun Qin. Transformed unscented Kalman filter. *IEEE Transactions on Automatic Control*, 58(1):252–257, 2013.

[28] S.Y. Chen. Kalman filter for robot vision: A survey. *IEEE Transactions on Industrial Electronics*, 59(11):4409–4420, 2012.

[29] J.M.C. Clark, R.B. Vinter, and M.M. Yaqoob. The shifted Rayleigh filter for bearings only tracking. In *Information Fusion, 2005 8th International Conference on*, volume 1, pages 8–16. IEEE, 2005.

[30] J.M.C. Clark, R.B. Vinter, and M.M. Yaqoob. Shifted Rayleigh filter: A new algorithm for bearings-only tracking. *IEEE Transactions on Aerospace and Electronic Systems*, 43(4), 2007.

[31] Pau Closas, Carles Fernandez-Prades, and Jordi Vila-Valls. Multiple quadrature Kalman filtering. *IEEE Transactions on Signal Processing*, 60(12):6125–6137, 2012.

[32] Ronald Cools and Philip Rabinowitz. Monomial cubature rules since Stroud: a compilation. *Journal of Computational and Applied Mathematics*, 48(3):309–326, 1993.

[33] J.L. Crassidis. Sigma-point Kalman filtering for integrated GPS and inertial navigation. *IEEE Transactions on Aerospace and Electronic Systems*, 42(2):750–756, 2006.

[34] D. Crisan and A. Doucet. A survey of convergence results on particle filtering methods for practitioners. *IEEE Transactions on Signal Processing*, 50(3):736–746, 2002.

[35] F. Daowang, L. Teng, and H.Z. Tao. Square-root second-order extended Kalman filter and its application in target motion analysis. *IET Radar, Sonar and Navigation*, 4:329–335, 2010.

[36] P.K. Dash and R.K. Mallick. Accurate tracking of harmonic signals in VSC-HVDC systems using pso based unscented transformation. *International Journal of Electrical Power & Energy Systems*, 33(7):1315–1325, 2011.

[37] F. Daum. Nonlinear filters - beyond the Kalman filter. *IEEE A&E Systems Magazine*, 20:57–69, 2005.

[38] Frederick E. Daum. Exact finite dimensional nonlinear filters. In *Decision and Control, 1985 24th IEEE Conference on*, volume 24, pages 1938–1945. IEEE, 1985.

[39] Randal Douc and Olivier Cappé. Comparison of resampling schemes for particle filtering. *Proceedings of the 4th International Symposium on Image and Signal Processing and Analysis*, pages 64–69. IEEE, 2005.

[40] Arnaud Doucet, Nando de Freitas, and Neil Gordon. Sequential Monte Carlo Methods in Practice, Springer 2001.

[41] Arnaud Doucet and Adam M. Johansen. A tutorial on particle filtering and smoothing: Fifteen years later. *Handbook of Nonlinear Filtering*, 12(656-704):3, 2009.

[42] P.N. Dwivedi, P. Bhale, and A. Bhattacharyya. Quick and accurate state estimation of RV from RF seeker measurements using EKF. In *Proceedings of 2006 AIAA Guidance, Navigation, and Control Conference*, 2006.

[43] Geir Evensen. The ensemble Kalman filter: Theoretical formulation and practical implementation. *Ocean Dynamics*, 53(4):343–367, 2003.

[44] Friedrich Faubel and Dietrich Klakow. Further improvement of the adaptive level of detail transform: Splitting in direction of the nonlinearity. In *EUSIPCO*, pages 850–854, 2010.

[45] Eli Fogel and Motti Gavish. Nth-order dynamics target observability from angle measurements. *IEEE Transactions on Aerospace and Electronic Systems*, 24(3):305–308, 1988.

[46] Paul Frogerais, Jean-Jacques Bellanger, and Lotfi Senhadji. Various ways to compute the continuous-discrete extended Kalman filter. *IEEE Transactions on Automatic Control*, 57(4):1000, 2012.

[47] A.T. Fuller. Analysis of nonlinear stochastic systems by means of the Fokker-Planck equation. *International Journal of Control*, 9(6):603–655, 1969.

[48] Valverde G. and V. Terzija. Unscented Kalman filter for power system dynamic state estimation. *IET Generation, Transmission & Distribution*, 5(1):29–37, 2011.

[49] Alan Genz. Numerical computation of multivariate normal probabilities. *Journal of Computational and Graphical Statistics*, 1(2):141–149, 1992.

[50] Alan Genz. Fully symmetric interpolatory rules for multiple integrals over hyper-spherical surfaces. *Journal of Computational and Applied Mathematics*, 157(1):187–195, 2003.

[51] Alan Genz and Bradley D. Keister. Fully symmetric interpolatory rules for multiple integrals over infinite regions with Gaussian weight. *Journal of Computational and Applied Mathematics*, 71(2):299–309, 1996.

[52] Alan Genz and John Monahan. A stochastic algorithm for high-dimensional integrals over unbounded regions with Gaussian weight. *Journal of Computational and Applied Mathematics*, 112(1):71–81, 1999.

[53] Thomas Gerstner and Michael Griebel. Numerical integration using sparse grids. *Numerical Algorithms*, 18(3-4):209, 1998.

[54] Thomas Gerstner and Michael Griebel. Dimension–adaptive tensor–product quadrature. *Computing*, 71(1):65–87, 2003.

[55] S. Gillijns, O.B. Mendoza, J. Chandrasekar, B.L.R. De Moor, D.S. Bernsein, and A. Ridley. What is the ensemble Kalman filter and how well does it work? In *Proceedings of the American Control Conference*, Minneapolis, USA, 2006. IEEE.

[56] Gene H. Golub and John H. Welsch. Calculation of Gauss quadrature rules. *Mathematics of Computation*, 23(106):221–230, 1969.

[57] Neil J. Gordon, David J. Salmond, and Adrian F.M. Smith. Novel approach to nonlinear/non-Gaussian Bayesian state estimation. In *IEE Proceedings F (Radar and Signal Processing)*, volume 140, pages 107–113. IET, 1993.

[58] N.J. Gordon, D.J. Salmond, and A.F.M. Smith. Novel approach to nonlinear/non-Gaussian state estimation. *IEE Proceedings-F*, 140(2):107–113, 1993.

[59] M.S. Grewal and A.P. Andrews. Applications of Kalman filtering in aerospace 1960 to the present. *IEEE Control Systems*, 30(3):69–78, 2010.

[60] Geoffrey Grimmett and David Stirzaker. *Probability and Random Processes*. Oxford University Press, 2004.

[61] Ravindra D. Gudi, Sirish L. Shah, and Murray R. Gray. Adaptive multirate state and parameter estimation strategies with application to a bioreactor. *AIChE Journal*, 41(11):2451–2464, 1995.

[62] F. Gustafsson, F. Gunnarsson, N. Bergman, U. Forsell, J. Jansson, R. Karlsson, and P.J. Nordlund. Particle filters for positioning, navigation, and tracking. *IEEE Transactions on Signal Processing*, 50(2):425–437, 2002.

[63] F. Gustafsson and G. Hendeby. Some relations between extended and unscented Kalman filters. *IEEE Transactions on Signal Processing*, 60(2):545–555, 2012.

[64] A.C. Harvey. *Forecasting, Structural Time Series Models and the Kalman Filter*. Cambridge University Press, 2008.

[65] Martin Havlicek, Karl J. Friston, Jiri Jan, Milan Brazdil, and Vince D. Calhoun. Dynamic modeling of neuronal responses in fMRI using cubature Kalman filtering. *NeuroImage*, 56(4):2109–2128, 2011.

[66] Florian Heiss and Viktor Winschel. Likelihood approximation by numerical integration on sparse grids. *Journal of Econometrics*, 144(1):62–80, 2008.

[67] A. Hermoso-Carazo and J. Linares-Pérez. Unscented filtering algorithm using two-step randomly delayed observations in nonlinear systems. *Applied Mathematical Modelling*, 33(9):3705–3717, 2009.

[68] Aurora Hermoso-Carazo and Josefa Linares-Pérez. Extended and unscented filtering algorithms using one-step randomly delayed observations. *Applied Mathematics and Computation*, 190(2):1375–1393, 2007.

[69] Francis Begnaud Hildebrand. *Introduction to Numerical Analysis*. Courier Corporation, 1987.

[70] R.A. Horn and C.R. Johnson. *Matrix Analysis*. Cambridge University Press, 1985.

[71] Jingyi Huang, Alex B. McBratney, Budiman Minasny, and John Triantafilis. Monitoring and modelling soil water dynamics using electromagnetic conductivity imaging and the ensemble Kalman filter. *Geoderma*, 285:76–93, 2017.

[72] Tahir Husain. Kalman filter estimation model in flood forecasting. *Advances in Water Resources*, 8(1):15–21, 1985.

[73] K. Ito and K. Xiong. Gaussian filters for nonlinear filtering problems. *IEEE Transactions on Automatic Control*, 45:910–927, 2000.

[74] Kazufumi Ito and Kaiqi Xiong. Gaussian filters for nonlinear filtering problems. *IEEE Transactions on Automatic Control*, 45(5):910–927, 2000.

[75] Claude Jauffret, Annie-Claude Pérez, and Denis Pillon. Observability: Range-only versus bearings-only target motion analysis when the observer maneuvers smoothly. *IEEE Transactions on Aerospace and Electronic Systems*, 53(6):2814–2832, 2017.

[76] Andrew H. Jazwinski. *Stochastic Processes and Filtering Theory*. Courier Corporation, 2007.

[77] Bin Jia and Ming Xin. Vision-based spacecraft relative navigation using sparse-grid quadrature filter. *IEEE Transactions on Control Systems Technology*, 21(5):1595–1606, 2013.

[78] Bin Jia and Ming Xin. Multiple sensor estimation using a new fifth-degree cubature information filter. *Transactions of the Institute of Measurement and Control*, 37(1):15–24, 2015.

[79] Bin Jia, Ming Xin, and Yang Cheng. Sparse Gauss-Hermite quadrature filter with application to spacecraft attitude estimation. *Journal of Guidance, Control, and Dynamics*, 34(2):367–379, 2011.

[80] Bin Jia, Ming Xin, and Yang Cheng. Sparse-grid quadrature nonlinear filtering. *Automatica*, 48(2):327–341, 2012.

[81] Bin Jia, Ming Xin, and Yang Cheng. High-degree cubature Kalman filter. *Automatica*, 49(2):510–518, 2013.

[82] Bin Jia, Ming Xin, and Yang Cheng. Relations between sparse-grid quadrature rule and spherical-radial cubature rule in nonlinear Gaussian estimation. *IEEE Transactions on Automatic Control*, 60(1):199–204, 2015.

[83] Simon J. Julier and Jeffrey K. Uhlmann. A new extension of the Kalman filter to nonlinear systems. In *Int. Symp. Aerospace/Defense Sensing, Simul. and Controls*, volume 3, pages 182–193. Orlando, FL, 1997.

[84] S.J. Julier. The spherical simplex unscented transformation. In *Proceedings of American Control Conference*, pages 2430–2434, Denver, USA, 2003. IEEE.

[85] Julier Simon J. and Jeffrey K. Uhlmann. New extension of the Kalman filter to nonlinear systems. In Signal processing, sensor fusion, and target recognition VI, vol. 3068, pp. 182–194. International Society for Optics and Photonics, 1997.

[86] S.J. Julier and J.K. Uhlmann. Reduced sigma point filters for the propagation of means and covariances through nonlinear transformations. In *Proceedings of American Control Conference*, pages 887–892, Anchorage, USA, 2002. IEEE.

[87] S.J. Julier and J.K. Uhlmann. Unscented filtering and nonlinear estimation. *Proceedings of the IEEE*, 92(3):401–422, 2004.

[88] Thomas Kailath. *Linear Systems*, volume 156. Prentice-Hall Englewood Cliffs, NJ, 1980.

[89] R.E. Kalman. A new approach to linear filtering and prediction problems. *Transactions of the ASME- Journal of Basic Engineering*, 82:35–45, 1960.

[90] R.E. Kalman and R.S. Bucy. New results in linear filtering and prediction theory. *Journal of Basic Engineering*, pages 95–108, 1961.

[91] C.D. Karlgaard and H. Schaub. Huber-based divided difference filtering. *Journal of Guidance, Control and Dynamics*, 30(3):885–891, 2007.

[92] Freda Kemp. An introduction to sequential Monte Carlo methods. *Journal of the Royal Statistical Society: Series D (The Statistician)*, 52(4):694–695, 2003.

[93] Maryam Kiani and Seid H. Pourtakdoust. Adaptive square-root cubature–quadrature Kalman particle filter for satellite attitude determination using vector observations. *Acta Astronautica*, 105(1):109–116, 2014.

[94] Genshiro Kitagawa. Monte Carlo filter and smoother for non-Gaussian nonlinear state space models. *Journal of Computational and Graphical Statistics*, 5(1):1–25, 1996.

[95] Jayesh H. Kotecha and Petar M. Djuric. Gaussian sum particle filtering. *IEEE Transactions on Signal Processing*, 51(10):2602–2612, 2003.

[96] Vladimir Ivanovich Krylov and Arthur H. Stroud. *Approximate Calculation of Integrals*. Courier Corporation, 2006.

[97] G. Yu Kulikov and Maria V. Kulikova. Accurate continuous–discrete unscented Kalman filtering for estimation of nonlinear continuous-time stochastic models in radar tracking. *Signal Processing*, 139:25–35, 2017.

[98] Kundan Kumar and Shovan Bhaumik. Higher degree cubature quadrature Kalman filter for randomly delayed measurements. In *2018 21st International Conference on Information Fusion (FUSION)*, pages 1589–1594. IEEE, 2018.

[99] Dirk Laurie. Calculation of Gauss-Kronrod quadrature rules. *Mathematics of Computation of the American Mathematical Society*, 66(219):1133–1145, 1997.

[100] Pei H. Leong, Sanjeev Arulampalam, Tharaka A. Lamahewa, and Thushara D. Abhayapala. A Gaussian-sum based cubature Kalman filter for bearings-only tracking. *IEEE Transactions on Aerospace and Electronic Systems*, 49(2):1161–1176, 2013.

[101] X. Rong Li and Vesselin P. Jilkov. Survey of maneuvering target tracking: III. measurement models. In *Signal and Data Processing of Small Targets 2001*, volume 4473, pages 423–447. International Society for Optics and Photonics, 2001.

[102] X. Rong Li and Vesselin P. Jilkov. Survey of maneuvering target tracking. part I. dynamic models. *IEEE Transactions on Aerospace and Electronic Systems*, 39(4):1333–1364, 2003.

[103] X. Rong Li and Zhanlue Zhao. Evaluation of estimation algorithms part I: incomprehensive measures of performance. *IEEE Transactions on Aerospace and Electronic Systems*, 42(4), 2006.

[104] Zhaoming Li and Wenge Yang. Spherical simplex-radial cubature quadrature Kalman filter. *Journal of Electrical and Computer Engineering*, 2017, 2017.

[105] Yan Liang, Tongwen Chen, and Quan Pan. Optimal linear state estimator with multiple packet dropouts. *IEEE Transactions on Automatic Control*, 55(6):1428–1433, 2010.

[106] Jun S. Liu and Rong Chen. Sequential Monte Carlo methods for dynamic systems. *Journal of the American Statistical Association*, 93(443):1032–1044, 1998.

[107] James Lu and David L. Darmofal. Higher-dimensional integration with Gaussian weight for applications in probabilistic design. *SIAM Journal on Scientific Computing*, 26(2):613–624, 2004.

[108] C.S. Manohar and D. Roy. Monte carlo filters for identification of nonlinear structural dynamical systems. *Sadhana*, 31(4):399–427, 2006.

[109] Raman Mehra. A comparison of several nonlinear filters for reentry vehicle tracking. *IEEE Transactions on Automatic Control*, 16(4):307–319, 1971.

[110] Dong Meng, Lingjuan Miao, Haijun Shao, and Jun Shen. A seventh-degree cubature Kalman filter. *Asian Journal of Control*, 2018.

[111] Jinhao Meng, Guangzhao Luo, and Fei Gao. Lithium polymer battery state-of-charge estimation based on adaptive unscented Kalman filter and support vector machine. *IEEE Transactions on Power Electronics*, 31(3):2226–2238, 2016.

[112] Qiang Miao, Lei Xie, Hengjuan Cui, Wei Liang, and Michael Pecht. Remaining useful life prediction of lithium-ion battery with unscented particle filter technique. *Microelectronics Reliability*, 53(6):805–810, 2013.

[113] Aström K.J. and Murray R.M. Feedback Systems: An Introduction for Scientists and Engineers. *Princeton University Press*; Apr 12, 2010.

[114] Darko Mušicki. Bearings only single-sensor target tracking using Gaussian mixtures. *Automatica*, 45(9):2088–2092, 2009.

[115] M.A. Myers and R.H. Luecke. Process control applications of an extended Kalman filter algorithm. *Computers & Chemical Engineering*, 15(12):853–857, 1991.

[116] Maciej Niedźwiecki and Piotr Kaczmarek. Estimation and tracking of complex-valued quasi-periodically varying systems. *Automatica*, 41(9):1503–1516, 2005.

[117] M. Norgaard, N.K. Poulsen, and O. Ravn. New developments in state estimation for nonlinear systems. *Automatica*, 36:1627–1638, 2000.

[118] Katsuhiko Ogata. *Modern Control Engineering*, volume 4. Prentice Hall, India, 2002.

[119] Kumar Pakki, Bharani Chandra, Da-Wei Gu, and Ian Postlethwaite. Cubature information filter and its applications. In *American Control Conference (ACC), 2011*, pages 3609–3614. IEEE, 2011.

[120] K. Ponomareva and P. Date. Higher order sigma point filter: A new heuristic for nonlinear time series filtering. *Applied Mathematics and Computation*, 221:662–671, 2013.

[121] William H. Press, Saul A. Teukolsky, William T. Vetterling, and Brian P. Flannery. *Numerical Recipes in C*, volume 2. Cambridge University Press Cambridge, 1996.

[122] Rahul Radhakrishnan, Shovan Bhaumik, and Nutan Kumar Tomar. Continuous-discrete shifted Rayleigh filter for underwater passive bearings-only target tracking. In *Control Conference (ASCC), 2017 11th Asian*, pages 795–800. IEEE, 2017.

[123] Rahul Radhakrishnan, Shovan Bhaumik, and Nutan Kumar Tomar. Continuous-discrete quadrature filters for intercepting a ballistic target on reentry using seeker measurements. *IFAC-PapersOnLine*, 51(1):383–388, 2018.

[124] Rahul Radhakrishnan, Shovan Bhaumik, and Nutan Kumar Tomar. Gaussian sum shifted Rayleigh filter for underwater bearings-only target tracking problems. *IEEE Journal of Oceanic Engineering*, (99):1–10, 2018.

[125] Rahul Radhakrishnan, Shovan Bhaumik, and Nutan Kumar Tomar. Continuous-discrete filters for bearings-only underwater target tracking problems. Early access *Asian Journal of Control*, 2019; 1–11.

[126] Rahul Radhakrishnan, Manika Saha, Shovan Bhaumik, and Nutan Kumar Tomar. Ballistic target tracking and its interception using suboptimal filters on reentry. In *2016 Sixth International Symposium on Embedded Computing and System Design (ISED)*, pages 274–278. IEEE, 2016.

[127] Rahul Radhakrishnan, Abhinoy Kumar Singh, Shovan Bhaumik, and Nutan Kumar Tomar. Multiple sparse-grid Gauss-Hermite filtering. *Applied Mathematical Modelling*, 40(7-8):4441–4450, 2016.

[128] Rahul Radhakrishnan, Ajay Yadav, Paresh Date, and Shovan Bhaumik. A new method for generating sigma points and weights for nonlinear filtering. *IEEE Control Systems Letters*, 2(3):519–524, 2018.

[129] A. Ray, L.W. Liou, and J.H. Shen. State estimation using randomly delayed measurements. *Journal of Dynamic Systems, Measurement, and Control*, 115(1):19–26, 1993.

[130] Branko Ristic, Sanjeev Arulampalam, and Neil Gordon. Beyond the Kalman filter. *IEEE Aerospace and Electronic Systems Magazine*, 19(7):37–38, 2004.

[131] Mohammad Rohmanuddin and Augie Widyotriatmo. A novel method of self-tuning PID control system based on time-averaged Kalman filter gain. In *Instrumentation Control and Automation (ICA), 2013 3rd International Conference on*, pages 25–30. IEEE, 2013.

[132] Anirban Roy and Debjani Mitra. Multi-target trackers using cubature Kalman filter for doppler radar tracking in clutter. *IET Signal Processing*, 10(8):888–901, 2016.

[133] Smita Sadhu, M. Srinivasan, Shovan Bhaumik, and Tapan Kumar Ghoshal. Central difference formulation of risk-sensitive filter. *IEEE Signal Processing Letters*, 14(6):421–424, 2007.

[134] Andrew P. Sage and James L. Melsa. Estimation theory with applications to communications and control. Technical report, Southern Methodist Univ Dallas Tex Information and Control Science Center, 1971.

[135] Rathinasamy Sakthivel, Subramaniam Selvi, Kalidass Mathiyalagan, and Peng Shi. Reliable mixed H_∞ and passivity-based control for fuzzy markovian switching systems with probabilistic time delays and actuator failures. *IEEE Transactions on Cybernetics*, 45(12):2720–2731, 2015.

[136] Simo Sarkka. On unscented Kalman filtering for state estimation of continuous-time nonlinear systems. *IEEE Transactions on Automatic Control*, 52(9):1631–1641, 2007.

[137] Simo Särkkä. *Bayesian Filtering and Smoothing*, volume 3. Cambridge University Press, 2013.

[138] Simo Särkkä and Arno Solin. On continuous-discrete cubature Kalman filtering. *IFAC Proceedings Volumes*, 45(16):1221–1226, 2012.

[139] A. Sharma, Suresh Chandra Srivastava, and Saikat Chakrabarti. A cubature Kalman filter based power system dynamic state estimator. *IEEE Transactions on Instrumentation and Measurement*, 66(8):2036–2045, 2017.

[140] Hermann Singer. Generalized Gauss-Hermite filtering. *AStA Advances in Statistical Analysis*, 92(2):179–195, 2008.

[141] Abhinoy Kumar Singh and Shovan Bhaumik. Quadrature filters for maneuvering target tracking. In *Recent Advances and Innovations in Engineering (ICRAIE), 2014*, pages 1–6. IEEE, 2014.

[142] Abhinoy Kumar Singh and Shovan Bhaumik. Higher degree cubature quadrature Kalman filter. *International Journal of Control, Automation and Systems*, 13(5):1097–1105, 2015.

[143] Abhinoy Kumar Singh, Shovan Bhaumik, and Paresh Date. Quadrature filters for one-step randomly delayed measurements. *Applied Mathematical Modelling*, 40(19-20):8296–8308, 2016.

[144] Abhinoy Kumar Singh, Paresh Date, and Shovan Bhaumik. A modified Bayesian filter for randomly delayed measurements. *IEEE Transactions on Automatic Control*, 62(1):419–424, 2017.

[145] Abhinoy Kumar Singh, Rahul Radhakrishnan, Shovan Bhaumik, and Paresh Date. Computationally efficient sparse-grid Gauss-Hermite filtering. In *Control Conference (ICC), 2017 Indian*, pages 78–81. IEEE, 2017.

[146] Bruno Sinopoli, Luca Schenato, Massimo Franceschetti, Kameshwar Poolla, Michael I Jordan, and Shankar S Sastry. Kalman filtering with intermittent observations. *IEEE Transactions on Automatic Control*, 49(9):1453–1464, 2004.

[147] George M. Siouris, Guanrong Chen, and Jianrong Wang. Tracking an incoming ballistic missile using an extended interval Kalman filter. *IEEE Transactions on Aerospace and Electronic Systems*, 33(1):232–240, 1997.

[148] Sergey Smolyak. Quadrature and interpolation formulas for tensor products of certain classes of functions. In *Soviet Math. Dokl.*, volume 4, pages 240–243, 1963.

[149] Halil Ersin Söken and Chingiz Hajiyev. UKF based in-flight calibration of magnetometers and rate gyros for pico satellite attitude determination. *Asian Journal of Control*, 14(3):707–715, 2012.

[150] Taek Lyul Song. Observability of target tracking with bearings-only measurements. *IEEE Transactions on Aerospace and Electronic Systems*, 32(4):1468–1472, 1996.

[151] Harold W. Sorenson and Daniel L. Alspach. Recursive Bayesian estimation using Gaussian sums. *Automatica*, 7(4):465–479, 1971.

[152] Harold W. Sorenson and Allen R. Stubberud. Non-linear filtering by approximation of the a posteriori density. *International Journal of Control*, 8(1):33–51, 1968.

[153] Jason L. Speyer and Walter H. Chung. *Stochastic Processes, Estimation, and Control*, volume 17, SIAM, 2008.

[154] Sarot Srang and Masaki Yamakita. Application of continuous-discrete unscented Kalman filter for control of nonlinear systems with actuator nonlinearity. In *Control Conference (CCC), 2014 33rd Chinese*, pages 8837–8842. IEEE, 2014.

[155] N.M. Steen, G.D. Byrne, and E.M. Gelbard. Gaussian quadratures for the integrals $\int_0^\infty exp(-x^2)f(x)dx$ and $\int_0^b exp(-x^2)f(x)dx$. *Mathematics of Computation*, 23(107):661–671, 1969.

[156] Srebra B. Stoyanova. Cubature formulae of the seventh degree of accuracy for the hypersphere. *Journal of Computational and Applied Mathematics*, 84(1):15–21, 1997.

[157] Ondřej Straka, Jindřich Duník, and Miroslav Šimandl. Gaussian sum unscented Kalman filter with adaptive scaling parameters. In *Information Fusion (FUSION), 2011 Proceedings of the 14th International Conference on*, pages 1–8. IEEE, 2011.

[158] N. Subrahmanya and Y.C. Shin. Adaptive divided difference filtering for simultaneous state and parameter estimation. *Automatica*, 45:1686–1693, 2009.

[159] Feng Sun and Li-Jun Tang. Cubature particle filter. *Systems Engineering and Electronics*, 33(11):2554–2557, 2011.

[160] Shuli Sun. Optimal linear filters for discrete-time systems with randomly delayed and lost measurements with/without time stamps. *IEEE Transactions on Automatic Control*, 58(6):1551–1556, 2013.

[161] Shuli Sun and Jing Ma. Linear estimation for networked control systems with random transmission delays and packet dropouts. *Information Sciences*, 269:349–365, 2014.

[162] Shuli Sun, Lihua Xie, and Wendong Xiao. Optimal full-order and reduced-order estimators for discrete-time systems with multiple packet dropouts. *IEEE Transactions on Signal Processing*, 56(8):4031–4038, 2008.

[163] Shuli Sun, Lihua Xie, Wendong Xiao, and Yeng Chai Soh. Optimal linear estimation for systems with multiple packet dropouts. *Automatica*, 44(5):1333–1342, 2008.

[164] Denis Talay. Numerical solution of stochastic differential equations. 1994. *Stochastics and Stochastic Reports*, 47(1–2):121–126, 1994.

[165] Xiaojun Tang, Jie Yan, and Dudu Zhong. Square-root sigma-point Kalman filtering for spacecraft relative navigation. *Acta Astronautica*, 66(5):704–713, 2010.

[166] D. Tenne and T. Singh. The higher order unscented filter. In *Proceedings of the American Control Conference*, Denver, USA, June 2003.

[167] Gabriel Terejanu, Puneet Singla, Tarunraj Singh, and Peter D. Scott. Adaptive Gaussian sum filter for nonlinear Bayesian estimation. *IEEE Transactions on Automatic Control*, 56(9):2151–2156, 2011.

[168] Lloyd N. Trefethen. Is Gauss quadrature better than Clenshaw-Curtis? *SIAM Review*, 50(1):67–87, 2008.

[169] Rudolph Van Der Merwe, Arnaud Doucet, Nando De Freitas, and Eric A. Wan. The unscented particle filter. In *Advances in Neural Information Processing Systems*, pages 584–590, 2001.

[170] Rudolph Van Der Merwe and Eric A. Wan. The square-root unscented Kalman filter for state and parameter-estimation. *IEEE International Conference on Acoustics, Speech, and Signal Processing, 2001. Proceedings (ICASSP'01)*, volume 6, pages 3461–3464, IEEE, 2001.

[171] Harry L. Van Trees. *Detection, Estimation, and Modulation Theory, Part I: Detection, Estimation, and Filtering Theory*. John Wiley & Sons, 2004.

[172] M. Verhaegen and P. Van Dooren. Numerical aspects of different Kalman filter implementations. *IEEE Transactions on Automatic Control*, 31(10):907–917, 1986.

[173] Dong Wang, Fangfang Yang, Kwok-Leung Tsui, Qiang Zhou, and Suk Joo Bae. Remaining useful life prediction of Lithium-ion batteries based on spherical cubature particle filter. *IEEE Transactions on Instrumentation and Measurement*, 65(6):1282–1291, 2016.

[174] Shiyuan Wang, Jiuchao Feng, and K Tse Chi. Spherical simplex-radial cubature Kalman filter. *IEEE Signal Processing Letters*, 21(1):43–46, 2014.

[175] Xiaoxu Wang, Yan Liang, Quan Pan, and Chunhui Zhao. Gaussian filter for nonlinear systems with one-step randomly delayed measurements. *Automatica*, 49(4):976–986, 2013.

[176] Xiaoxu Wang, Yan Liang, Quan Pan, Chunhui Zhao, and Feng Yang. Design and implementation of Gaussian filter for nonlinear system with randomly delayed measurements and correlated noises. *Applied Mathematics and Computation*, 232:1011–1024, 2014.

[177] Grzegorz W. Wasilkowski and Henryk Wozniakowski. Explicit cost bounds of algorithms for multivariate tensor product problems. *Journal of Complexity*, 11(1):1–56, 1995.

[178] Curt Wells. *The Kalman Filter in Finance*, volume 32. Springer Science & Business Media, 2013.

[179] Yuanxin Wu, Dewen Hu, Meiping Wu, and Xiaoping Hu. A numerical-integration perspective on Gaussian filters. *IEEE Transactions on Signal Processing*, 54(8):2910–2921, 2006.

[180] Lihua Xie, Yeng Chai Soh, and Carlos E. De Souza. Robust Kalman filtering for uncertain discrete-time systems. *IEEE Transactions on Automatic Control*, 39(6):1310–1314, 1994.

[181] Paul Zarchan. *Tactical and Strategic Missile Guidance*. American Institute of Aeronautics and Astronautics, Inc., 2012.

[182] Huanshui Zhang, Gang Feng, and Chunyan Han. Linear estimation for random delay systems. *Systems & Control Letters*, 60(7):450–459, 2011.

[183] Ling Zhang and Xian-Da Zhang. An optimal filtering algorithm for systems with multiplicative/additive noises. *IEEE Signal Processing Letters*, 14(7):469–472, 2007.

[184] Xin-Chun Zhang. Cubature information filters using high-degree and embedded cubature rules. *Circuits, Systems, and Signal Processing*, 33(6):1799–1818, 2014.

[185] Yonggang Zhang, Yulong Huang, Ning Li, and Zhemin Wu. Sins initial alignment based on fifth-degree cubature Kalman filter. In *Mechatronics and Automation (ICMA), 2013 IEEE International Conference on*, pages 401–406. IEEE, 2013.

[186] Yonggang Zhang, Yulong Huang, Zhemin Wu, and Ning Li. Seventh-degree spherical simplex-radial cubature Kalman filter. In *Control Conference (CCC), 2014 33rd Chinese*, pages 2513–2517. IEEE, 2014.

Index

air traffic control, 23
aircraft tracking, 105
anti-ballistic missile, 216
autopilot, 217

ballistic coefficient, 219, 223
ballistic missile, 216
Bayes' rule, 9
bearing density, 194
Bernoulli random variable, 141
Brownian motions, 160

central difference approximation, 45
Chapman-Kolmogorov equation, 10
Chebyshev-Laguerre polynomial, 68
Cholesky decomposition, 47, 52
constant velocity model, 189
continuous-discrete system, 168
        nonlinear, 169
coordinated turn model, 106
cross covariance, 21
cubature Kalman filter, 68
cubature quadrature points, 70
curse of dimensionality problem,
        87, 99

delta function, 119
diffusion matrices, 160
direction cosine matrix, 222
discretization, 172
drag force, 223

ensembles, 23
ergodicity, 34
Euler method, 173
extended Kalman filter, 38
extended Kalman-Bucy filter, 167

filtering, 7
Fokker-Planck equation, 171

Gauss-Hermite filter, 96
Gauss-Laguerre integration rule, 68
Gauss-Laguerre quadrature, 79
Gaussian filter, 17
        assumption, 20
        measurement update, 21
        time update, 20
Gaussian sum approximation, 118
Gaussian sum shifted Rayleigh filter,
        193
GPS-INS, 217
guidance
        homing, 216
        initial, 216
        mid-course, 216
        proportional navigation, 217,
        225
        terminal, 216

Hessian, 45

iid, 12
information matrix, 164
innovation, 165
interceptor, 217
Ito-Taylor expansion, 173

Jacobian, 38, 51, 170
Joseph form, 32

Kalman filter, 27
        application, 23
        convergence, 31
        stability, 31
Kalman gain, 30, 142

Kalman-Bucy filter, 161
Kolmogorov's forward equation, 170

likelihood, 10
linear time invariant, 3

maneuvering target, 106
Markov process, 6
Maximum a posteriori estimate, 9
maximum likelihood, 8
mean square error, 161
measurement update
    Gauss-Hermite filter, 112
    filtering with randomly delayed
        measurement, 153
    Gaussian sum filter, 123
    Kalman filter, 30
    unscented Kalman filter, 55
miss-distance, 217
moment matching, 53, 104
multiple sparse-grid GHF, 109

navigation, 24
NEES, 23
new unscented Kalman filter, 61
non-additive noise, 55

observability Gramian, 165

particle filter, 12
    particle impoverishment, 16
    weight update, 14
    importance sampling, 12
    proposal density, 14
    resampling, 15
    systematic resampling, 16
    weight degeneracy, 15
    weight normalization, 14
passive sensor measurement, 188
point mass description of a pdf, 12
posterior pdf, 9
product rule, 97, 100
prediction, 7

QR decomposition, 75
quaternion, 222

RADAR target tracking
    continuous discrete filter, 179
randomly delayed measurement
    any step, 146
    single step, 143
Riccati equation
    algebraic, 165
    differential, 163
    discrete algebraic, 31
RMSE, 22, 201
Runge-Kutta method, 172

seeker, 220
sequentian importance sampling,
    12
shifted Rayleigh filter, 191
sigma point
    new sigma point filter, 63
sigma point filter, 49
sigma points, 52
six DOF, 219
Smolyak's rule, 100
smoothing, 7
sparse-grid GHF, 99
spherical cubature rule, 66
    higher degree, 78
    third degree, 67
spherical simplex transformation,
    60
square root
    cubature quadrature Kalman
        filter, 75
state partitioning, 109
state space
    continuous time, 2
    discrete time, 3
state transition matrix, 165
stochastic differential equation,
    160
subspace, 109
support points, 54

target motion analysis, 188
target tracking, 23
Taylor series, 38, 45

time to go, 226
time update
　filtering with randomly delayed
　　measurement, 153
　Gaussian sum filter, 122
　Kalman filter, 29
　unscented Kalman filter, 55
track-loss, 201

tracking with RADAR, 129
tridiagonal matrix, 96

unscented Kalman filter, 54

weight normalization, 121
weight update
　adaptive Gaussian sum, 124
　Gaussian sum filter, 123